THE CAPE CO
NATIONAL
SEASHORE

Experiments in Bioregionalism
The New England River Basins Story
Charles H. W. Foster, 1984

The Cape Cod National Seashore
A Landmark Alliance
Charles H. W. Foster, 1985

THE CAPE COD NATIONAL SEASHORE

A Landmark Alliance

Charles H. W. Foster

Published for Tufts University by
University Press of New England
HANOVER AND LONDON, 1985

University Press of New England

Brandeis University

Brown University

Clark University

University of Connecticut

Dartmouth College

University of New Hampshire

University of Rhode Island

Tufts University

University of Vermont

Printed in the United States of America

LIBRARY OF CONGRESS CATALOGING-IN-PUBLICATION DATA

Foster, Charles H.W., 1927–
 The Cape Cod National Seashore.

 Bibliography: p.
 Includes index.
 1. Cape Cod National Seashore (Mass.)—History
2. United States. Cape Cod National Seashore Advisory
Commission. 3. Coastal zone management—
Massachusetts
—Cape Cod. I. Title.
F72.C3F67 1985 974.4′92 84–40583
ISBN 0–87451–346–4 (pbk.)

Map on p. xiv is reprinted, with permission, from Francis P.
Burling, *The Birth of the Cape Cod National Seashore,* Leyden
Press, 1979.

Contents

Preface

This volume is the second in a set of case studies concerned with the various institutions that have been created to manage bioregional resources in the six states that comprise New England, where bioregional experimentation has been in full flower since the early days of the colonies.

Bioregionalism is the act of responding to natural resources or environmental issues and events that occur in transboundary settings. Good examples are river systems that flow indiscriminately across state and even international boundaries, migratory birds and fish that range on a continental or transoceanic scale, and regionwide environmental effects such as the movement of contaminants through air, water, or living systems. If natural resources are to be managed as whole systems as the ecologists suggest, then techniques for the management of particular resources in particular places must be adapted to the transboundary settings in which they occur naturally. And all of this must take place within a political environment generally hostile to institutions extending beyond conventional boundaries.

The setting for this case study is Cape Cod, Massachusetts. Unlike the earlier volume, which dealt with twenty-eight defined river basins within a fifty-thousand square mile region extending from the Canadian border to Long Island Sound, this Cape Cod bioregion embraces a modest fifty square miles of land and water and thirty-nine miles of ocean beach along what is inelegantly referred to as the "backside" of Cape Cod. Rather than six states, there are six New England towns—Provincetown, Truro, Wellfleet, Eastham, Orleans, and Chatham—independent communities incorporated more than two hundred years ago that, to this very day, defend fiercely their individual rights and prerogatives. Into this inhospitable environment in 1961 came a new national park—the Cape Cod National Seashore—the first of a new generation of national conservation projects and a pioneer in several important respects. Contained within the authorizing legislation for the Seashore was provision for a statutory advisory commission consisting of representatives of the jurisdictions affected.

The pages that follow recount the experience of the Cape Cod National Seashore Advisory Commission during the first twenty years of its existence. The account is written from my perspective as a participant in the early days and an interested observer in later years. The record has been drawn from the minutes of 143 official meetings (nearly four thousand pages in all) and from personal interviews with more than sixty of its participants. At the end of this account, an attempt is made to analyze that experience and relate it beyond the purely Cape Cod setting to the larger question of how any conservation area can be properly responsive to the localities within which it occurs, yet also faithful to the mission it represents. But a second experience is recounted as well—how the Seashore communities were able to overcome their innate parochialism and perform a distinguished role in the management of a significant bioregional resource.

In principle, six unifying elements tend to contribute to a viable bioregion.[1] First it must be *spatially* distinct, marked by such elements as topography, landscape, climate, soils, drainage, or biota. It must also have a sense of *social* identity (i.e., be popularly identifiable as a region). An *economic* identity is usually helpful—a recognized place to produce goods, deliver services, and sustain commercial activities. Ready definition in *planning* or *administrative* terms can provide two other potentially unifying elements. Finally, a sense of *political* identity and cohesion is important, for a bioregion must inevitably have the capacity to advance its own interests and resolve at least a measure of its own problems. Not each of the six unifying elements is of equal significance, however. If the area in question does not have a high degree of social and political integrity, any bioregional program or institution is unlikely to prevail. Conversely, the more unifying elements that overlap and are mutually reinforcing, the more likely it is that a bioregional program or institution will succeed. Prevailing biological wisdom notwithstanding, mere ecological integrity will not be persuasive unless people also see themselves as living in a distinct region.

In the Seashore's case, the Great Beach and its associated uplands provide a distinct spatial identity. Less certain are its other environments—the heathlands, bay shore, and inland marshes—which extend beyond the Seashore's authorized boundaries. The management of these resources requires cooperative action by other jurisdictions and, as such, stimulates the Seashore's participation in larger bioregional efforts.

As a social concept, Cape Cod has always been distinct. In art, history, literature, and culture, it is widely known. Within the region itself, the lower Cape is recognized as different. Its social organizations are often distinctive. It has its own newspapers and television station.

In economic terms, there is no dominant industry. Traditional agricultural and fishing activities have declined over the years, and manufacturing is limited to local crafts. The largest economic activities are land development and service to seasonal visitors. Of emerging significance is Cape Cod's position as a retirement community, a growth industry of its own. Delineated by the transportation "spine" of Routes 6 and 28, the lower Cape has an economic identity, but its real economic center is the Seashore itself.

Viewed administratively and politically, the dimensions of the bioregion are at least marginally distinct. Administrative and planning activities tend to concern themselves with Barnstable County as a whole, whose limits extend even beyond the Cape Cod Canal—the natural boundary for Cape Cod. But even for Cape-wide projects, such as the master plan prepared by the Cape Cod Planning and Economic Development Commission, it is not unusual to treat the lower Cape as a separate subregion. The six towns involved in the Seashore represent distinct political units, although their small, year-round populations give them little political clout in either the state capital or in Washington.

Turning to the bioregional institutions themselves, the prognosis for most is still far from favorable, in the New England region at least. We suspect that institutional viability is associated in some way with the size, degree of representativeness, active participation, and continuity of its membership. The manner of creation and the provisions for accountability are other important factors. Regional acceptance can also hinge upon the size and scope of the operation and the degree of professionalism displayed by the management and staff. Authority questions invariably arise. There is a mistaken tendency to equate influence with authority. In the particular quicksand of multiple political jurisdictions, the influence and, indeed, the survival of a bioregional institution may actually be favored by minimal authority. Powerless to compete against and threaten other jurisdictions, the institution's views, oddly, may gain special credence. The tendency to regard bioregional institutions as fixed and stable entities ignores the reality that their operating environments are changing constantly. Thus, the institutions with the built-in flexibility to adjust their activities, modify their membership, and remain truly responsive to the concerns of their constituencies, are those most likely to succeed over time.

The sum total of the above suggests that the Cape Cod National Seashore, and its Advisory Commission, came into being within a setting innately capable of sustaining a viable, bioregional program. Whether or not this promise has been fulfilled will be discussed in the account that follows.

Acknowledgments

The idea for this book grew out of the author's personal service from 1962 to 1966 as the first chairman of the Cape Cod National Seashore Advisory Commission. The concept crystallized in 1975 when the Conservation Foundation, a Washington-based policy and research center with special interests in parks and recreation, agreed to sponsor an evaluation of the Cape Cod experience. The resources for the study, limited to expense reimbursement only, were derived privately. When the work began, two other participants were recruited to lend a hand. One was Francis P. Burling, former managing editor of *The Cape Codder*, the leading weekly newspaper for the lower Cape; the other was Robert F. Gibbs, the man who served as the first superintendent of the Cape Cod National Seashore. Both individuals came out of retirement to constitute the project's interview team and to develop materials for which they had special insight or expertise. Just before his untimely death, Burling published separately his recollections under the title *The Birth of the Cape Cod National Seashore* (Leyden Press, Plymouth, Mass.), describing in human terms the period of legislative enactment which he had covered so extensively for *The Cape Codder.* Gibbs's account of the effort to acquire the lands and waters of the Seashore and to administer its special zoning provisions is the cornerstone of the descriptive and land acquisition sections of the present book. The intent of the study team was to capture not just the facts but the full flavor of the Cape Cod experience before time erased such memories. It is literally impossible to credit the many individuals who made that possible, but a few should be mentioned at this time.

William K. Reilly and William E. Shands of the Conservation Foundation provided encouragement throughout the duration of the project. The five superintendents of the Seashore and the four Advisory Commission chairmen during this period were liberal with their impressions and insights. Valuable clerical and support services were provided by the Bio/Enviro Section of Arthur D. Little Inc. of Cambridge, Massachusetts. Lastly, special thanks are owed to Marjorie S. Burling (Mrs. Francis P. Burling) of Brewster, who willingly and capably transcribed the interview material, allowed the special map of Cape Cod to be reproduced, and, from her spe-

cial perspective as personal secretary to four of the Seashore's superintendents, reinforced our sense of a project worth doing. It is to her that this account is gratefully dedicated.

Needham, Massachusetts   C. H. W. F.
December 1984

THE CAPE COD
NATIONAL
SEASHORE

Introduction

Fifty miles southeast of Boston, as a determined sea gull might fly, lies the northernmost tip of Cape Cod. Thence thirty-nine miles in a southerly direction extends the area known as the Great Beach, the longest stretch of unspoiled seashore on America's entire North Atlantic coast. Yet, paradoxically, Cape Cod represents significant human as well as natural history. It was Bartholomew Gosnold, for example, arriving in the region on May 14, 1603, who conducted the first explorations and named the new land for the schools of fish so plentiful off its shores. The famed Mayflower Compact was signed not in Plymouth but in Cape Cod's Provincetown Harbor on November 11, 1620. And as the settlers came ashore to find their first fresh water (Pilgrim Spring), their first food (Corn Hill), and their first glimpse of Indians (First Encounter Beach), it was foreordained that the Cape Cod story would be one of close interaction between man and land. Small wonder then that National Park Service director Conrad Wirth encountered not representatives of Cape Cod's nine Indian sachems, but of six militantly independent New England towns, when he took the podium at Eastham's crowded town hall the evening of March 23, 1959, to announce that his agency would actively pursue the establishment of a national seashore park on Cape Cod.[2]

The story of the Seashore's establishment has been ably told by Francis P. Burling in his insider's account of the legislative proceedings entitled *The Birth of the Cape Cod National Seashore.* Suffice it to say, almost two years would elapse between the submission of Senate 2636, the initial legislation drafted by Massachusetts senators Leverett Saltonstall and John F. Kennedy, and the enactment of Public Law 87-126, to authorize the establishment of the Cape Cod National Seashore, on August 7, 1961. To prepare the reader for the rapid pace of people, places, and events to follow, this brief natural and social history of Cape Cod has been prepared.[3]

Like so much of New England, the landscape of Cape Cod is a reflection of its geological history. During the final stages of the Pleistocene period fifteen thousand years ago, glacial ice covered all of Cape Cod. When the ice began to melt, the unsorted mass of materials (till) transported by the glacier formed ridges known as moraines. One such formation, the Sandwich moraine, is the dominant landscape form of Cape Cod. But the glacier

1

contributed other features as well. It left behind occasional large boulders (erratics), such as Doane Rock in Eastham, which bear witness to the geological events of the past. Blocks of glacial ice, embedded in the ground, were replaced by rising groundwater to form the rounded, "kettle-hole" ponds of Wellfleet and Truro so favored by summer residents and visitors. And the glacier created so-called outwash plains—deposits of gravel, silt, and clay formed by the glacial meltwater streams as they discharged to the ocean. These served as the earliest sites for agriculture and remain important today as promising sources of fresh water.

But the landscape of Cape Cod is also the by-product of the constant interaction of the sea and its adjacent land mass. At roughly the midpoint of the Great Beach in Truro, Wellfleet, and Eastham, the high cliffs described so eloquently by Henry David Thoreau continue to erode at a rate of three feet a year. Longitudinal ocean currents transport eroded material north and south to nourish the barrier beaches at Provincetown and Chatham. Farther inland, the process of accretion has created connected land masses of what are still termed Great, Griffin, and Boundbrook islands by native Wellfleeters and the official U.S. Geological Survey. Astonishing though it may seem today, ships' owners used to watch from these sites for their fleets at sea. In earlier times, the actions of wind and sea are believed to have created an open passage from ocean to bay in the vicinity of Eastham; it is now marked indistinctly by a series of shallow ponds and inland marshes. Eastham's Salt Pond estuary and Orleans's Pleasant Bay are reminders of the power of these past interactions. Even today, the ocean periodically flexes its awesome muscles. The great storm of February 1978, for example, breached the beaches in Eastham and Provincetown in a matter of hours.

At least seven distinct landscape types occur within the Cape Cod National Seashore. The *Great Beach* is its most dramatic feature, extending from Race Point, Provincetown, to the Monomoy National Wildlife Refuge in Chatham. In Provincetown and Truro, the beach is wide, gently sloping, and backed by high dunes. From Highland Light south to Coast Guard Beach in Eastham, the beach narrows below high sand cliffs to provide an unusual sense of isolation. South of Coast Guard Beach, the beach takes the form of a sand spit less than a quarter of a mile wide backed by low dunes, salt marshes, and, still farther south, the open waters of Pleasant Bay. An important characteristic of the Great Beach is its convexity. In a sizable portion of the Seashore, a user can enjoy the ocean beach with minimal visual intrusion from others. Behind the Seashore in Provincetown and Truro lie some eight square miles of spectacular *dunes*, some more than eighty feet in height. The impact is especially dramatic coming into Provincetown on Route 6, where Pilgrim Lake serves as a reflecting body for a particularly massive set of migrating dunes. The scene is a dynamic one — sunken forests engulfed by sand, areas of tangled shrubs and waving beach grass, and patches of delicate and beautiful wildflowers. In South Wellfleet and North Eastham, a relatively level *plain* extends from Route 6 to the edge of the sea cliffs, the site of small farms until their abandonment and subsequent invasion by pitch pine, scrub oak, and beach plum. Moving southward, another dominant landscape feature is *Nauset Marsh*. The marsh itself is extensive with abundant wildlife and distinctive ecology; but the setting is noteworthy as well, for the marsh is surrounded by rolling hills and old fields invaded by juniper, a panorama readily appreciated from the Salt Pond Visitor Center in Eastham. Outward from the Great Beach, enhanced by the bars and rips created by constant wave action against the shore, lies a rich and productive *offshore* environment, one of the ten best saltwater sport fishing areas in the United States.

Within the inland portions of the towns of Truro and Wellfleet, the landscape shifts again. Ancient river valleys and quiet, deep-set, freshwater *ponds* provide a striking contrast to the restless Atlantic. Here also occur the lovely "pamets," the Indian term for the stream-cut, outwash channels that juxtapose heath, marsh, and water so strikingly within these winding inland valleys. And on the *Cape Cod Bay* side of the Seashore occurs an entirely different landscape, even more serene in nature. Thirteen additional miles of gently sloping beach surround Jeremy Point, the narrow peninsula that separates Wellfleet Harbor from Cape Cod Bay. Extensive fresh water marshes adjoin the Herring River and the former "islands." A striking feature is the heath-covered hilltops, interspersed with old homes that command a panoramic view of the harbors and surrounding woodlands. It is here that traditional Cape Cod architecture is at its classic best.

Associated with these landscapes are important plant and animal communities. Except for small remnants of the original beech and maple forest, the vegetation is now dominantly pitch pine and scrub oak, often the aftermath of wildfires and agricultural land abandonment. Yet remarkable species diversity still exists within the Seashore, including extensive areas of bayberry barrens, salt grass, and beach grass communities. Thirty-six species of mammals are known to inhabit these areas and, at times of peak migration, more than 150 species of birds have been identified in a single day. Saltwater fish of many species, including shellfish, are taken both by commercial and recreational fishermen. Small but active fishing fleets ply their trade from Provincetown and Chatham, and the recreational boating industry is extensive throughout the entire lower Cape. But it was Henry Beston, writing in *The Outermost House*, who described the setting best:

"outermost cliff and solitary dune, the plain of the ocean and the far, bright rims of the world, meadow land and marsh and ancient moor: this is the. . . outer Cape".[4]

Viewed culturally, Cape Cod's earliest inhabitants were, of course, Indians, who were a dominant force until their instrument of submission to King James in 1621. After annexation by the Plymouth Colony in 1630, Cape Cod began the process of colonization and settlement characteristic of all the colonies. Beginning in 1640, the colonial legislature (termed the Great and General Court) followed the practice of awarding grants of land to small groups of individuals who wished to live and worship together. The official business of the community came to be dealt with at annual town meetings where each inhabitant was accorded the right to speak and vote on a given issue. Between such sessions, a smaller group of chosen individuals, termed *selectmen*, conducted the business of the community, holding office on an elective basis. Selectmen exercised only those powers granted to them by town meeting action; they were merely the spokesmen and executive officers for their communities. This remarkable form of direct democracy remains the practice in all of Cape Cod today.[5]

In the early days, sheer subsistence was the dominant imperative. The early settlers were primarily farmers and herdsmen who found Cape Cod's bony soil hardly hospitable to agriculture. Grain crops, such as wheat, corn, oats, and rye, were grown as food for both man and animal. Marsh grass was harvested as hay for livestock. A new agricultural industry – the growing of cranberries – had its origin on Cape Cod. Taught by friendly Indians, the early settlers also became proficient at fishing, learning to harvest the abundant finfish and shellfish in Cape Cod waters. A British naval base established in Provincetown during the Revolutionary War interrupted this promising activity, necessitating a rebuilding of the fishing fleet when hostilities ceased, a precursor of the extensive development investments Boston merchants would make in later years. At the turn of the nineteenth century, the first saltworks were installed on Cape Cod, an industry that enjoyed only marginal success. It was not until the advent of the railroad, the installation of major new road systems such as the mid-Cape highway, and the construction of the Cape Cod Canal, that transportation improvements would stimulate the Cape's most significant industry – land development and visitor services.

By the mid-twentieth century, the outer region of Cape Cod was a thriving, self-contained, political community composed of six highly independent towns. Its year-round population was twelve thousand, swelling at least fivefold during the summer months.[6] Thanks to a heavy reliance on the seasonal visitor and a minimal demand for year-round services, the communities were remarkably well-off. Land was plentiful, the demand for vacation properties was high and growing, and the local governance system was such that the town official could play a direct role in the development process. Indeed, many were personally involved—as landowners, facility operators, or purveyors of essential legal, financial, appraisal, engineering, or building services. Planning was primitive at best; zoning was in its infancy. To many on the outside and an increasing number of native Cape Codders, the region was threatening to kill the goose that was laying the golden egg—tourist revenues—through uncontrolled development. The situation was especially rampant in Wellfleet and Truro, and increasingly so in the seashore portions of Eastham. The northernmost town, Provincetown, hemmed in by the state-owned Province Lands, was appreciably different. Its extensive colony of artists and artesans made it the intellectual capital of the lower Cape. At the other end, Chatham's relative sophistication and its primary orientation toward the Nantucket Sound side of Cape Cod made the problem somewhat less acute here.

But an odd coalition of summer residents and year-rounders was beginning to form—a coalition dedicated to the proposition that some sort of major conservation action was essential to the future well-being of the Cape. Town and state action would not be enough; a national park for Cape Cod appeared both timely and appropriate. The early advocates of this approach were not popular, but their stature was such as to command thoughtful attention. In the Massachusetts legislature was Edward C. Stone, a summer resident of Hyannis and the senior and respected state senator representing the Cape and the islands. In government service was Francis W. Sargent, Massachusetts commissioner of natural resources, himself a summer resident of Orleans. At the local level was John Worthington, local businessman and selectman from the town of Truro. And in Congress, the unlikely but eminently workable combination of Republican senator Leverett Saltonstall and Democratic senator John F. Kennedy had privately instructed their personal staffs to make the preservation of the lower Cape and, particularly, its Great Beach a matter for priority attention.[7]

Wirth's appearance in Eastham's town hall on the moonlit night of March 23, 1959, was premature,[8] because the Seashore's enabling legislation would not be entered into Congress until September 3, 1959. But as the National Park Service director told some five hundred assembled Cape Codders, the agency's survey report, supported privately by foundation contributions, had recommended a national park to include both the Great Beach and an extensive inland acreage stretching clear across the Cape to the bay shore of Wellfleet. It was this issue more than any other that stirred local sensibilities, spurring Rev. Earl B. Luscombe, pastor of the Wellfleet and Eastham Methodist churches, to declaim: "Men of the lower Cape, come, meet the foe, cast off all lethargy; arise to meet the present hour!"[9] This local interests did—at field hearings in Eastham in December of 1959 and 1960, and at additional legislative hearings in Washington during the early part of 1961. But by August of 1961, with cosponsor John F. Kennedy newly installed as the nation's thirty-fifth president, the legislation emerged from conference committee with a favorable recommendation. It was signed into law by President Kennedy on August 7, 1961.

As enacted by the Eighty-seventh Congress, Public Law 87-126 establishing the Cape Cod National Seashore was a lengthy document.[10] Virtually one-third of its provisions specified the exact boundaries of the new Seashore. Another third detailed the landmark provisions for land acquisition and local zoning. A final substantive section of the bill provided for a Cape Cod National Seashore Advisory Commission, an instrument that carried forward Senator John F. Kennedy's earlier observations of an effort to "properly harmonize the national, state and private interests which are involved in a venture of this nature."[11]

Section 8 of the act established an advisory commission of ten members, appointed by the secretary of the interior from nominations submitted by eight specified jurisdictions. Each of the six lower Cape communities was allotted a representative, as was Barnstable County as a whole. Two members would represent the commonwealth of Massachusetts. One additional member would be appointed directly by the secretary of the interior, who would also designate the chairman of the Commission. Advisory Commission members were to serve terms of two years' duration but were subject to renomination without limit by their respective jurisdictions. The Commission as a whole would terminate ten years after the official establishment of the Seashore by the secretary.

The act was silent on the exact manner of operation of the Advisory Commission except that it would operate through majority vote. Congress purposely avoided rigid detail in specifying the responsibilities of the Commission, preferring a large measure of judgment and discretion on the part of participants.

After much debate about giving the Commission actual powers, the act included a general section 8(f) calling for consultation with the Commission "from time to time" with respect to development and acquisition matters, and a specific section 8(g) requiring the secretary of the interior to seek the advice of the Commission before issuing any permits for commercial or industrial uses of property or establishing any public-use areas for recreational activity. The statute specified that the Advisory Commission must submit such advice "within a reasonable time after it is sought."

Although much of the earlier debate over the Seashore legislation had focused on its size and boundaries, the advisory commission proposal had also attracted attention. The issues were those of numbers, representation, powers, and duration. The earlier legislative drafts had called for a permanent advisory commission of nine members with purely advisory powers. After the field hearings in Eastham in December of 1959 and 1960,[12] Senate sponsors had incorporated a Barnstable County representative, language specifying the commission's role in commercial and recreational uses, and a ten-year life for the Commission.[13]

Congressman Hastings Keith of the Massachusetts Ninth District was the chief champion of the Advisory Commission,[14] but if the National Park Service were to have its way, no advisory commission would be created. Director Conrad Wirth's personal attitude toward such advisory bodies, dating back to his father's experience as superintendent of parks for the city of Minneapolis, had encouraged a servicewide policy of opposition to the establishment of park advisory bodies.[15]

Nevertheless, by November of 1961 Acting Assistant Director Robert W. Ludden could advise Wirth that nominations for chairman of the Advisory Commission had been received from all of the governmental groups specified in the legislation.[16] Despite a request for two nominees, from which the secretary of the interior would select one member, most of the jurisdictions had nominated a single individual to be certain that the secretary had no such latitude. There was lively competition for the slot of secretary's designee with the matter ultimately settled by President Kennedy himself.[17]

In a prudent move, Director Wirth was actively advocating as chairman one of the governor's nominees, Massachusetts commissioner of natural resources Charles H. W. Foster. The Park Service had worked with Foster in the past in various professional capacities, whereas the secretary of the interior's nominee was certain to be a political appointment. Besides, a state member was apt to sit better with Cape Cod residents than a representative of the Washington bureaucracy.

By early January, official notices of appointment were in the mail to the new Advisory Commission members, requesting acceptances "in view of your interest in conservation on Cape Cod."[18] The official Interior press announcement went out on January 9, 1962.[19] It was assumed that the first meeting of the Advisory Commission would be held within a matter of weeks, and on Cape Cod. There was, therefore, widespread reaction and considerable speculation when the first meeting date was announced for February 16—in Washington, D.C.

The Cape Cod
Advisory Commission

THE FIRST MEETING

Ominous skies greeted the Cape Cod contingent as the newly appointed Advisory Commission headed for its first meeting in the nation's capital. With snow on the ground, Provincetown's representative Nathan Malchman debated with his wife about the wisdom of driving to Boston's Logan Airport. She lost and he went.[20] The Truro, Eastham, and Barnstable County representatives joined forces for the trip to Washington. Charles Foster, the newly appointed chairman, decided to take no chances, riding the overnight train from Back Bay to Union Station with a fresh copy of *Robert's Rules of Order* in hand. The Federal took this occasion to be two hours' late due to a derailment between Boston and Providence. Nevertheless, the bulk of the Commission was on hand when the session convened at 10:00 A.M. on Wednesday, February 16, 1962, in room 5160 of the Interior building—the secretary of the interior's opulent conference room.[21] Like many similar events, much had already transpired before the opening gavel was struck.

The National Park Service's northeast regional director Ronald F. Lee had been placed in charge of the event by his Washington office.[22] He had conferred with Foster about meeting arrangements. Foster had suggested having the first meeting in Washington to supply a clear sense of direction to the Commission.[23] All subsequent sessions should be held on Cape Cod, he advised. Teletyped invitations were in the mail to participants when a draft agenda was sent by Lee to Foster and Director Conrad Wirth on February 2. The proposed business was divided into two parts: matters relating to the organization of the Commission and status reports on the Seashore from various Park Service officials. Regrettably, Secretary Stewart Udall would be occupied that day, but Assistant Secretary Carver would do the welcoming honors in his stead. A full day's worth of business had been scheduled for the fledgling Commission.

On the Cape, the preparations were less distinguished but equally thorough. Secretary Udall's appointments to the Commission on January 8 had

9

raised some eyebrows. As the *Boston Globe*'s Cape Cod correspondent reported,[24] four of the ten Advisory Commission members were strong opponents of the park. One of the four, Barnstable County businessman Joshua A. Nickerson, had arranged an informal caucus of the lower Cape representatives in Orleans on February 9 to develop a strategy for the Washington meeting. Commonwealth representative Josiah H. Child of Provincetown, an ardent supporter of the Seashore from the outset, remembers a jaundiced eye being cast on his presence at the meeting.[25] Another observer felt it singularly high-handed for Nickerson to have begun to formulate a subcommittee or bloc within the Commission without the chairman's prior knowledge.[26] Nevertheless, the local members of the Commission were agreed on six items of business to transact at the Washington meeting regardless of what the Park Service might have up its sleeve.[27]

The Cape Cod National Seashore Advisory Commission was, at best, a mixed lot when it assembled around the gleaming conference table that morning for its first meeting. A cynic might have suspected a deliberate effort to impress the Commission with the strength and majesty of the federal presence. The chairs were richly upholstered and a magnificent Moran portrait of the Grand Canyon hung on the wall.[28]

The Commission included the secretary of the interior's representative Leo E. Diehl, head of the Excise Tax Division of the Massachusetts Department of Corporations and Taxation, and a close associate of Massachusetts congressman Thomas P. O'Neill; two nominees of the governor of Massachusetts: Charles H.W. Foster, the commonwealth's young commissioner of natural resources, and Josiah H. Child, European-trained architect and Provincetown innkeeper; Joshua A. Nickerson of Harwich, a lumber company executive who served as the representative of the Barnstable County Commissioners; and six town nominees: Provincetown merchant Nathan Malchman, chairman of the town finance committee; selectman and attorney John R. Dyer, Jr., of Truro; former school teacher Esther Wiles of Wellfleet; former Eastham selectman and real estate agent Ralph A. Chase; Selectman Arthur Finlay of Orleans; and Selectman Robert A. McNeece of Chatham, owner of a small house-painting and decorating business and former sales manager of a wholesale crockery and glassware firm.

In the absence of the chairman, Director Conrad Wirth of the National Park Service assumed the chair.[29] Neither Secretary Udall nor Assistant Secretary Carver were available, he told the Commission, but Secretary Udall did expect to join the group for lunch. Administrative Assistant Secretary D. Otis Beasley was then pressed into service to present the official greetings of the department. But he did more than that. To the utter astonishment of the Commission and other individuals present, he proceeded to read at length from the department's manual on the functions of advisory committees. In short, the Cape Cod National Seashore Advisory Commission's role would be solely advisory; action would be determined only by a department official; all meeting arrangements and agenda items would require prior approval; meetings could only be conducted in the presence of a federal officer who would have authority to adjourn the session whenever he felt the public interest was not being served. To the Commission, it was just what many had feared.

Nickerson was the first to pick up the gauntlet. What about the specific language of the Cape Cod Act, he inquired? It appeared to conflict with the procedures contained in the manual. The law will prevail, answered Beasley. Other Commission members chimed in: What does the act mean by "reasonable time" for Commission advice. How is "fair market value" to be determined? What will the procedures be in the case of town-owned lands?

Soft-spoken McNeece gave his opinion that the session should be confined to organizational matters only. The questions and answers could come later. Possibly, Nickerson growled, but taking away the authority of the Commission to have its own meetings is a poor beginning.

Beasley took advantage of the momentary lull to hand the gavel back to Director Wirth, who went on at some length to describe the background and mission of the National Park Service and its reporting and operating channels. Wirth closed with a fervent plea for a spirit of friendship and cooperation to prevail. Wellfleet's Mrs. Wiles was not moved. One can promote good feeling only with justice, she snapped.

Assistant Director Hillory A. Tolson stepped into the breach by distributing annual passes to the Commission members "as a token of appreciation of your coming down here." At this point Chairman Foster walked into the room and the gavel passed hands once again.[30] After a few introductory remarks, the organizational business of the Commission was underway.

McNeece of Chatham was persuaded to accept the post of secretary after reassurances that he would be put to no undue time or expense. The departmental manual notwithstanding, the Commission determined that a vice-chairman would be in order and proceeded to elect Nickerson to that office. Introductions were made: George H. Thompson, the newly designated land acquisition officer for Cape Cod, and superintendent-designate Robert F. Gibbs of the Cape Hatteras National Seashore, termed the "captain of our ship" by an expansive Wirth. Gibbs, from a spectator seat at the side of the room, could not help but wonder how well his ship would sail with a crew like this.[31]

With some relief, Director Wirth suggested that the meeting be adjourned for lunch. The Cosmos Club, Washington's well-known mecca for the intellectual elite, had set aside a private dining room for Wirth's Commission guests. They were joined by a special group from Capitol Hill: Massachusetts senator Benjamin Smith, Representatives Hastings Keith and Edward Boland of the Massachusetts House delegation, and William Saltonstall, representing his father, Massachusetts senior senator Leverett Saltonstall, who could not be present.[32] In informal remarks following the luncheon, Interior secretary Stewart Udall assured the Advisory Commission that his agency would never "do injury to the people of the Cape." In turn, Wirth told his congressional guests that he was "extremely pleased with the attitude and help the Commission has shown." Not to be outdone, Chairman Foster characterized the session as "an excellent meeting with a capital *E.*" Even Vice-Chairman Nickerson reported that the National Park Service gave "eager attention" to the point of view and problems presented by Commission members.[33] Later, however, Foster expressed his personal feelings to Regional Director Ronald Lee on the "sour note" sounded by the Washington session. The Commission should be permitted to meet as it wishes, he said. "It is far preferable, in my judgment, to have a strong feeling of *voluntary,* mutual cooperation."[34]

THE EARLY YEARS

On Friday, March 9, 1962, the Advisory Commission began what was to become a long and illustrious career on Cape Cod. After its moment of elegance in the Washington sun, the Commission was now faced with reality. The plush chairs and ornate environment of the secretary of the interior's conference room had given way to hardback chairs and a green steel conference table pried loose from the General Services Administration in Boston by Assistant Regional Director Carlisle Crouch.[35]

The contentious Commission had been relegated appropriately to the basement of the former Nauset Life Boat Station in Eastham. Difficult though it was for some members to negotiate the thirteen narrow stairs leading to the basement conference room, the Commission seemed more at home as it crowded around the narrow conference table for its second official meeting.[36]

The Seashore's first superintendent, Robert Franklin Gibbs, was there to greet them. A Virginian by birth, Gibbs had come to Cape Cod by way of the Cape Hatteras National Seashore where he had also served as superintendent. Gibbs was big and plain spoken. There was a touch of the southern sun in his opening remarks to the Commission. The Commission liked him immediately.

There is evidence that considerable care was taken in the selection of the first personnel for the Seashore. Gibbs had been called to the phone late one night by National Park Service director Conrad Wirth. Wirth made it plain that Gibbs's presence on Cape Cod was a matter of some importance to the Service. Although such reassignments were technically a matter for concurrence by the superintendent, a Park Service professional did not turn down many such requests if he intended to continue to advance within the system.

Gibbs's first official exposure to the Seashore had not been a pleasant one. He had moved his family and possessions to a rented house only to find the house posted by the owner against his occupancy. Despite its apprehensions about the Seashore, the community reacted indignantly to this act of unneighborliness, rallying with offers of food and shelter. If Cape Cod expected fair play from the Seashore, it should be prepared to practice what it preached.[37]

Gibbs was also fortunate in finding Land Acquisition Officer George H. Thompson already on the scene. Unlike Gibbs, Thompson was familiar with the Cape. He had visited his sister in Falmouth many times and had been called upon to guide National Park Service officials during their various visits to the Cape. Thompson had been trained as a landscape architect at the University of Massachusetts but had served more recently as land acquisition officer for both the Midwest and Northeast Regional offices. Thompson had been called into Regional Director Tobin's office in Philadelphia one morning and given the welcome news of his assignment to Cape Cod.

Arriving to open the first Seashore office at the borrowed Coast Guard facilities in Eastham, Thompson had been confronted with a line of forty people looking for information about employment possibilities and land acquisition. A low-key, soft-spoken professional, Thompson had spent the bulk of his first four months on the job attending local meetings and putting to rest inaccurate reports of the Seashore's intentions.[38]

Some confusion had also arisen over just who would be Interior's official representative at Advisory Commission meetings. A Northeast Field Committee was in existence in Boston under the direction of Mark Abelson. The purpose of the Field Committee was to coordinate the activities of the various agencies and bureaus on a department basis. It was customary for the Northeast Field Committee to be used for advisory assignments in the region. There were some advantages to this arrangement since the director of the Northeast Field Committee reported directly to the Program Planning Staff of the secretary of the interior. In addition, Abelson and Chairman Foster were old friends and colleagues from water resources and river basin planning enterprises.

However, after strong lobbying by the National Park Service director, an agreement was reached at the secretarial level that Superintendent Gibbs would continue to be the official departmental officer at all meetings of the Advisory Commission.[39]

With real feeling, Gibbs told the Commission that Cape Cod was the "friendliest place" he had ever seen. The Commission responded by giving Gibbs a copy of *Nature's Year: The Seasons of Cape Cod* written by naturalist John Hay of the Cape Cod Museum of Natural History. With these auspicious beginnings, the Commission plunged into the work at hand.[40]

The Commission began by sharpening the procedures discussed at its first meeting in Washington. It was agreed that agenda items would be essentially predetermined, that the chairman would call all meetings, and that the Commission would be concerned primarily with policy matters. It confirmed the practice of executing its statutory approvals only at a subsequent meeting. From the outset, Chairman Foster required a degree of formality during the meeting, recognizing speakers officially and referring to them by their last names. Unconsciously, the Commission fell into the seating pattern first established in Washington. For every meeting of the Commission thereafter, the vice-chairman sat at the left of the chairman, and the secretary sat at his right.[41]

One other tradition emerged, probably as an outgrowth of the Commission's difficult session in Washington. From the second meeting on, no Park Service official was ever seated with the Commission at the conference table without prior invitation. Washington regulations notwithstanding, the Commission's business was its own. Government was merely an invited guest.[42]

The Commission then considered its external responsibilities. It was determined that while members were nominees of their respective jurisdictions, they were, in essence, United States advisory commissioners and should regard themselves as such at all times. A request from consulting master planner Van Ness Bates to meet the Advisory Commission resulted in a denial and a decision to deal primarily with local elected officials. In terms of press relations, the Commission determined to adopt the practice of holding a press conference immediately after each meeting in order to advise the public promptly of its discussions and recommendations.

As the day's first substantive agenda item, Camp Wellfleet, was debated, it was Eastham representative Ralph Chase who first suggested morning field trips to places of concern within the Seashore. Whole-day meetings of the Commission thereupon became the standard practice. Members of the press were invited to participate in the morning field trips in order to supplement the official accounts of Commission proceedings by informal discussions with Commission members. By these procedures, the Advisory

Commission was able to circumvent the current closed meeting policy for all federal advisory commissions and to ensure a steady flow of information to the press and to the public.[43]

The Commission was also confronted with a Presidential Executive Order relating to conflicts of interest.[44] Virtually all of the local members were in a potential conflict situation. Many had been or were still active in real estate either as agents or attorneys, for land and development were the primary growth industries of the Cape. In addition, several of the Commission members were the direct owners of prospective Seashore property. A delicate balance had to be struck between the intimate knowledge of property considerations the Advisory Commission could supply and the obvious conflict of interest their involvement would appear to create.

High on the Commission's priority list was the need to provide accurate public information soon about the status of the Seashore. The extensive national publicity about the Seashore legislation would certainly bring hordes of visitors in the summer expecting a National Park fully acquired and in operation. It was important that the public be informed that the Seashore was still essentially in private hands.

A prospectus on public information, prepared by the regional office of the National Park Service, was presented to the Advisory Commission. Its stated purpose was to prevent the inundation of the Seashore by uninformed visitors. Recommended was an information-interpretive station, a simple information folder depicting the broad outlines of the authorized Seashore, and a strongly worded Interior press release describing the present availability of public-use facilities at Cape Cod.[45] Provincetown businessman Nathan Malchman agreed with the approach philosophically, but urged that the Seashore not overdo its discouragement of summer visitors.[46]

Upon further exploration, a visitor station on Route 6 just before the Orleans traffic circle seemed to be the most appropriate location. A Cape Cod cottage was purchased and installed at roadside with the permission of the State Department of Public Works. Nearly five thousand inquiries were recorded at the station during the period July 8 through September 3, 1962. Fully one-third of the inquiries related to the availability of federal beach facilities.

Superintendent Gibbs also instituted a monthly news letter on Seashore affairs which was reprinted in its entirety by *The Cape Codder* and other local newspapers. Many years later, Director Wirth offered the observation that the newsletter had been a key factor in winning local acceptance of the Seashore.[47]

A second matter of pressing concern was the status of former Camp Wellfleet. Learning of its likely inactivation by the military, the National Park Service had instituted condemnation proceedings shortly after the Seashore Act was passed. It did so without detailed knowledge of the ownerships and without the statutory concurrence of the Advisory Commission or the town. Eighteen hundred acres in extent, the tract was just too critical to be lost through a technicality.

More important than the propriety of the taking was the condition of the land itself. It had been used for many years as a training ground for military operations, and much of the area was littered with live ammunition. The problem of decontamination was a massive one, described by military officials as likely to cost more than the land was worth. Yet there was unanimous agreement by the National Park Service and the Advisory Commission that the decontamination should proceed. After much correspondence at the departmental level between the departments of the Interior and the Army,[48] the commandant at Fort Devens gave the order to proceed with "neutralization." By the end of the year decontamination was complete. The Commission could now wrestle with the problems of what to do with the land, how the many buildings should be utilized, and what kind of development should take place.

It was Eastham representative Ralph Chase who first raised the issue of facility development.[49] What about the new bathhouse at Camp Wellfleet reported in the *Cape Cod Standard Times,* he asked? Could such a facility be built by the National Park Service without first seeking the advice of the Advisory Commission under section 8(g) of the act? Indeed it could not. The regional director of the Park Service, Ronald F. Lee, spoke up. In the future, the Advisory Commission would be involved intimately in every phase of the planning and development program.

True to his word, Lee brought to the next meeting Eugene DeSilets, supervising landscape architect of the Park Service's eastern design office. DeSilets was to become a steady visitor to Cape Cod as the Advisory Commission discussed in minute detail his policies and plans. On one occasion, DeSilets remarked that he now knew how it must feel to be sentenced more than once to the electric chair.[50]

DeSilets began by expressing the six basic policies of the Seashore master plan.[51] Point 4, "Keep to a minimum the disruption of the life of each town," struck a particularly responsive chord with the Commission. As Lee reported later to Director Wirth, the Commission found these planning principles to be basically acceptable.

In terms of development itself, however, the Commission had a good deal more to say. Strong interest was expressed in the preservation of marshes and freshwater bodies. The need for conservation and interpretation of historic features drew unanimous acclaim. Interior representative Leo Diehl wondered about the extent of concessions and marinas. Chairman Foster's comment favoring the consideration of recreation facilities in a regional context triggered a spirited discussion of campground use and development. This subject will require careful consideration, Ronald Lee later wrote in a summary memorandum to the director.[52]

But it was commonwealth representative Josiah H. Child who was to sound the theme that became the underlying principle of Seashore development policy and a rallying cry on many a subsequent occasion. We should make it plain, he said, whether the basic purpose of the Seashore is one of recreation or conservation. Barnstable County representative Joshua A. Nickerson turned to the language of the act. Congress clearly intended the Seashore to preserve the significant natural and historic values of Cape Cod, he said, and that is what our policy should be. The Commission was in unanimous accord.[53]

The stage 1 master plan in its present form was clearly unacceptable and must be rewritten. Under the circumstances, its title page depicting the Mayflower under full sail in heavy seas seemed particularly appropriate.

By the end of the first year, a much-revised master plan was back before the Commission and generally acceptable. Having learned its lessons the hard way, the National Park Service also brought before the Commission its initial development package containing twelve projects. The fiscal 1963 budget had produced $400,000 for initial construction, and the prospects for fiscal 1964 appeared equally promising. The Commission gave its blessing to these endeavors without delay.

On the land acquisition side, there was also much to be done. Here the Seashore already had a good start, for Land Acquisition Officer George Thompson had arrived on the Cape in October of 1961, had recruited a small staff, and had been authorized $4 million for land acquisition during the first year. Yet, since this was the Service's first major land acquisition effort in history, progress was tortuous at best. Thompson encountered delay after delay in securing approvals for perimeter surveys, title examiners, and appraisers. Many letters were being received from interested sellers, but procedures and policies were still far from fully developed. Thompson recalled the early days as hectic ones marked by endless telephone calls, local meetings, and press inquiries.[54]

In discussing the land acquisition situation, the Advisory Commission urged special priority for hardship cases. Beyond that, it felt that the larger tracts should be approached first, particularly where sensitive resources were involved. And where possible, the purchases should be made so that whole areas could be blocked into federal ownership.

The Seashore's land acquisition program was delayed by still another concern—the absolute conviction of Gibbs and Thompson that the government's land acquisition program should be as sensitive as possible to landowner concerns. In that respect, George Thompson was the ideal person for a major land acquisition effort involving a series of settled communities and unsettled local owners. Property owners were astonished by the patient and courteous manner of Thompson's approach. He left behind a growing legacy of respect and confidence in the Service's land acquisition program. Perhaps the "Great White Father" was not to be as feared as many had earlier expected.[55]

Of particular concern to the Advisory Commission was the manner in which town-owned properties would be handled. Just how would the secretary of the interior go about seeking the town concurrence specified in the act? The possible fate of the many miles of town beach, a source of great local pride to the lower Cape communities, was uppermost in mind. An official inquiry directed to the secretary brought back a logical response: a "mutually satisfactory formula" would be worked out with the Commission. Seashore Superintendent Gibbs and the Advisory Commission did just that, drafting guidelines for the secretary of the interior as well as suggested procedures for the towns to follow should the matter of federal acquisition of municipal property come up for consideration. The Department of the Interior accepted the Commission's recommendations with only minor changes.[56]

Two other land acquisition matters were of particular concern to the Commission. The first involved the practice of seasonal cottage rentals, a tradition for many year-round Cape Cod residents. Would these be regarded as commercial enterprises. The answer was no.[57]

The second matter was potentially more serious. The Seashore Act of August 7, 1961, carried provisions for a one-year moratorium on the exercise of the secretary of the interior's condemnation authority, allowing the towns a reasonable period of time to enact the zoning bylaws that would suspend the secretary's authority permanently for improved properties within the Seashore. Unaware of the rigors of the New England town meeting system, Congress had not allowed sufficient time for the bylaws to be drafted, reviewed, and approved. Under these circumstances, would the secretary agree to withhold the use of his condemnation authority as long as the towns were making good-faith efforts to comply with the provisions of the act? He would.[58]

By the end of 1962, Land Acquisition Officer George Thompson could report that the program was beginning to move. The process of executing and receiving approvals for options was now in working order, and the first actual land purchase had been consummated. The Seashore had moved promptly to forestall a proposed development in Eastham by using the threat of condemnation. When its bluff was actually called by a developer in Wellfleet, the Seashore had responded promptly with condemnation proceedings. The lower Cape was astonished that government could move so rapidly, and this fact had a remarkable effect on other builders who had planned to test the development control features of the Act.[59]

Encouraging though these developments appeared to be, the real key to the establishment of a viable Seashore lay in the acquisition of enough land to permit operation and development. The early drafts of the Seashore legislation had contained a provision for official establishment after a minimum of six thousand acres had been acquired. The final version had deleted all acreage provisions. However, the Seashore would not be established officially by the secretary of the interior until, in his judgment, there was sufficient land for a viable project. The matter of legal establishment had little real significance in terms of programs and operation, but it did represent an important benchmark locally. The ten-year statutory life of the Advisory Commission would begin on the date of official establishment. The clock would begin ticking at that time.

State officials were thus urged to proceed with the promised transfer of the Province Lands and Pilgrim Spring State Park.[60] This contribution of more than five thousand acres by the commonwealth had been a dramatic part of the earlier legislative proceedings. The promise had been made by Governor Foster Furcolo and confirmed by his successor, Governor John A. Volpe, but legislative action would still be required. Outside of the tactical advantages of making such a commitment, there were practical benefits to be obtained from a transfer of the state acreage.

Under commonwealth ownership, the Province Lands were the administrative responsibility of the Department of Public Works, whereas Pilgrim Spring State Park was under the jurisdiction of the Department of Natural Resources. Transfer of these properties to the federal government would not only achieve economies in operation, but also ensure management under single jurisdiction. Yet the Province Lands, in particular, represented a sensitive issue for the commonwealth and the lower Cape because of their great antiquity and historical significance. While most would agree that the commonwealth had not done a good job in managing these lands, a state legislature was not ordinarily inclined to turn over property to the federal government without some sort of quid pro quo.

Chairman Foster, the state commissioner of natural resources, was determined to make good his promises at the earlier congressional hearings. With the full consent of Governor Volpe, a former commissioner of the Department of Public Works, transfer legislation was drafted and submitted to the legislature.[61] Pragmatic Cape Cod, however, was not going to let this issue go without trying to apply some sort of leverage on the federal government.

Accordingly, the Barnstable County commissioners established a six-member committee under the chairmanship of Joshua A. Nickerson which urged the legislature to postpone action on the transfer bill until concessions could be obtained from the federal government.[62] This position was countered by the argument that the transfer was the key to the official establishment of the Seashore. The decisive moment came the day before the Fourth of July when, at a public hearing of the Joint Legislative Committee on Harbors and Public Lands, the venerable and powerful Cape Cod senator Edward C. Stone spoke for nearly an hour spelling out the pros and cons of the issue. At the end of his discourse, and to the immense relief ot the proponents, Stone recommended that the transfer legislation be passed.[63]

The Barnstable County special committee did not return to the Cape entirely empty-handed. A companion bill had been enacted that reserved from the transfer for a period of fifteen year's time a forty-acre parcel of land in Provincetown for the construction of a marina facility. Proponents of the transfer legislation were willing to run this risk, because the citizen's Emergency Committee for the Province Lands, in conjunction with a special committee of the Highland Fish and Game Association, was prepared to watchdog the town of Provincetown and prevent any initiation of the marina project.

A curious feature of the transfer legislation was that it went only part way in transferring state-owned lands to the Seashore. Included in the legislative authorization was title to all of the natural ponds within the towns of Provincetown and Truro, but not the Great Ponds lying within the town of Wellfleet—bodies of water that were an integral part of the landscape the Seashore was designed to protect. Similarly, the commonwealth ceded ownership of tidelands to a distance of one-quarter mile from shore within the Provincetown-Truro area, but not the twelve thousand acres of tidelands lying offshore along the remainder of the Seashore.[64]

One other important matter had to be dealt with during this first year of Commission activity. The centerpiece of the Seashore legislation was its provision for the adoption of protective zoning as a precursor to the suspension of condemnation for improved properties. The towns needed to get on with the business of adopting approvable bylaws, a process normally fraught with hazard in the New England town meeting tradition.

The Advisory Commission determined that the zoning provisions were so important that a special effort should be made to assist the towns in their formulation and adoption. Upon request, Attorney-Planner Elmer V. Buschman of the Park Service's Washington office was assigned to the Cape Cod project.[65] Louis H. Smith, the chief planning engineer of the Massachusetts Department of Commerce, also was made available.[66] They were joined as a team by a staff member of Blair Associates, the planning firm hired by several of the towns individually and by Barnstable County for the preparation of a regional master plan.

The Buschman-Smith-Blair team met with planning board and public officials in each community on repeated occasions. Numerous drafts of bylaws were prepared. Because of the disparity of the individual town zoning bylaws, it became evident that each community would have to be approached individually and the Seashore zoning provisions tailor-made to fit the existing body of town law.

For strong Baptist Buschman, Cape Cod was a memorable experience. New England Unitarianism and the searching examination of issues on a community-wide basis were new experiences for him, as was also the thicket of town procedures and bylaws.[67] But both he and Smith were just right for the job—low key, thoughtful, yet tactfully persistent. The town of Truro, surprisingly, became the first community to adopt an approvable bylaw; the town of Wellfleet, predictably, the last. Yet, the ultimate town meeting actions—were virtually unanimous—a tribute to the hard work and many community meetings attended by Buschman and Smith, but also to the wisdom of the zoning provisions in the Seashore Act.

It is important to remember that the adoption of adequate zoning bylaws had important advantages to all participants: protection from condemnation for the landowner, preservation of an important tax base for the community, and reduction of land purchase expenses for the federal government. That the bylaws were admirably fitted to the communities and the Seashore is attested to by the fact that in thirteen years' time, there has yet to be an amendment to a Seashore bylaw. And less than forty variance requests from the existing bylaws have been submitted within the six towns encompassed by the Seashore.

Despite the volume of business at hand, the Advisory Commission took time to make a trip to the Cape Hatteras National Seashore late in 1962 for a firsthand view of what a national seashore was apt to be like. The representatives of all six towns made that trip.[68] For many, it was their first look at a national park area. The members returned reassured by the evidence of professionalism exhibited within the areas they visited.[69] Their discussions with counterpart local officials were equally encouraging. Less assuring, however, was their first glimpse at Park Service development, notably the public camp grounds at Cape Hatteras. The Advisory Commission returned with the conviction that Cape Cod could and should do better.

Early in 1963, the Advisory Commission began to wrestle with the fate of the more than seventy-five commercial establishments in existence within the boundaries of the Seashore. These ran the gamut from medical, legal, and real estate offices to gas stations and food service establishments. The Commission was on the horns of a dilemma. On the one hand, the Seashore's primary objectives were conservation and preservation. On the other hand, the many visitors to the Seashore would require at least a reasonable level of commercial services.

The nature of the lower Cape communities added still another dimension to the problem, for many of the regular community commercial services were located within the Seashore's boundaries. However, under section 8(g) of the act, the Advisory Commission was required to render advice to the National Park Service on the issuance of all commercial permits.

After much debate, the Commission decided that the only sensible procedure would be to issue a blanket, one-year continuance for all existing commercial operations. Thereafter, following case-by-case analysis, acceptable facilities would be granted five-year permits. For those facilities not likely to be continued or renewed, notification would be furnished at least three years in advance of the termination date.[70]

The Advisory Commission was confronted with a particularly difficult problem in the case of the Salt Pond and Nauset Knoll motels in Eastham and Orleans. These facilities were situated at strategic entrances to the Seashore. Since it was doubtful that the facilities could be expanded, their owners were anxious to sell. Federal ownership would provide absolute safeguards for the Seashore, but it also raised the next question: Should the motels be continued as federal concessions? At least one member of the Advisory Commission, Esther Wiles of Wellfleet, had expressed the conviction that it was morally wrong for the federal government to acquire commercial properties. She would feel doubly indignant if the federal government thereupon became the operator of the commercial facility.

Superintendent Gibbs had additional difficulties. The acquisition decision had to be made immediately, yet the Advisory Commission was hosting its counterpart body from the Minuteman National Historical Park in a joint session fully open to the public and the press.[71] Gibbs did not want this sensitive matter broached publicly before being taken up officially with the Advisory Commission.[72]

When the Commission finally did consider the matter of the two motels, the acquisition commitment had been made. The only question before the Commission was whether the facilities should be removed or operated under concession by the National Park Service. The Commission was clearly divided on the issue, but in the end Yankee frugality prevailed. At the urging of the Eastham and Orleans selectmen, the Advisory Commission voted to recommend the issuance of a concession prospectus for not more than five years' additional operation, at the end of which time both sets of motel facilities would be removed. It also favored a determined effort by the National Park Service to find some way of returning a measure of the concession revenues to the town as reimbursements in lieu of taxes.[73]

During 1963 the Advisory Commission also devoted time to fine tuning the master plan.[74] Many questions and suggestions were raised but no serious conflicts emerged. The Commission appeared to accept at face value the prevailing concept of an evolving master plan. Individual members spoke of the need for a circulatory system for oversand vehicles and for greater recognition of the role of art and literature within the Seashore's interpretive program. The basic mission of preservation was enunciated time and time again, but questions were also raised as to the adequacy of the planned recreation facilities. Sharp local eyes spotted important defects in the master plan document, such as the absence of picnic areas in the Wellfleet portion, lack of consideration of wilderness, and the absence of any reference to the signing of the Pilgrim compact in Provincetown Harbor. The Advisory Commission was also told that famous Boston heart surgeon Paul Dudley White had been given news to "warm his own heart"—an Interior commitment to develop a network of bicycle trails within the Seashore.[75]

At the Commission's urging, the draft master plan was reviewed thoroughly by state officials, particularly the park and recreation planners of the Department of Natural Resources who were engaged in a simultaneous project to develop a state recreation plan for Barnstable County. To state recreation planner Lewis A. Carter, the Seashore's policies and ideas seemed exactly right. They promised to capture the "sights, sounds, and sensations" of the Seashore, he wrote in a complimentary staff report to his superiors.[76]

But for architect-member Josiah H. Child, the developer of award-winning plans for state beach facilities at Scusset, Horseneck, and Salisbury, the real test would be how the developed facilities were fitted to the land. Child was active in many arenas, fighting the battle of marshland preservation in Provincetown, yet also utilizing the prestige of his Advisory Commission appointment to help influence state and national policy. Acting on his own initiative, but with the full blessings of the chairman and the superintendent, Child composed a letter to twelve of the most distinguished architects in the nation, soliciting their concept of the design policies that should guide the Seashore's development program. Buttressed by these responses, and notably the one received from Walter Gropius, dean of the Harvard Graduate School of Design, the National Park Service was encouraged to develop an architectural theme for Seashore facilities that combined the best of the contemporary and traditional worlds. This encouragement beyond a pedestrian architectural approach would later come in handy as the Advisory Commission considered specific plans for bathhouse and visitor center buildings that were a far cry from conventional Cape Cod architecture.[77]

For fiscal 1964 the Seashore had been allotted more than $1.5 million for construction, the major components of which were to be the first visitor center, destined for a site overlooking Salt Pond in Eastham, and a complex of headquarter facilities which were to be located at former Camp Wellfleet. The first year's construction budget also included funds for an initial beach development project in Eastham if the town proceeded with its then-current plans to turn over Nauset Light and Coast Guard beaches to the Seashore for operation and development. True to his promise, Superintendent Gibbs brought the preliminary plans for every major development to the Advisory Commission for discussion and approval.[78]

By the summer of 1963, word of the new Seashore had traveled widely beyond Cape Cod. Visitation was picking up despite the absence of completed facilities. The first interpretive programs of the Seashore had proved to be outstanding successes. For one evening program, held inside the small auditorium at Camp Wellfleet, an overflow crowd upended trash barrels outside the windows for at least a glance at the interpretive slide show.[79]

A first set of Seashore-use regulations was drafted and reviewed thoroughly by state, local, and Advisory Commission officials. There were special use problems at Cape Cod which the National Park Service had never before encountered, such as the statutory provisions for continued hunting and fishing consistent with the regulations of the towns and the state. The Seashore's management was in a particular spotlight with the entire Cape Cod area beset by problems of juvenile rowdyism. There was both the necessity and the opportunity to develop a cooperative protection plan between the Seashore ranger force and the community and county police agencies.[80]

At the March 1963 town meeting in Eastham, the community voted to proceed with the transfer of town beaches to the Seashore. Some discussion about the loss of local rights occurred, but the issue of reduced operating expenses for the town was the factor that carried the day. A reverter clause for noncompliance by the federal government and a guarantee of free use of federal beaches by town residents satisfied all but die-hard opponents. The Eastham beaches had borne the full brunt of the Atlantic Ocean and were in need of substantial repair. As one former town official told the town meeting: "The federal government can't do worse than we have."[81]

A similar move was under way in neighboring Truro. Many of the town officials and a sizable number of town residents were in favor of federal assumption of beach responsibility. Yet as Truro representative John Dyer reported to the Advisory Commission,[82] there would be no transfer of town beaches unless some reimbursement for town expenditures could be worked out. Since the Seashore Act specifically precluded the acquisition of town property by anything but gift, the matter of the Truro town beaches was not likely to be settled in the near future.

Superintendent Gibbs was not entirely unhappy. He made it plain to the Advisory Commission that the federal government was not soliciting transfers at this time. With a new Seashore on his hands and the first federal beach now assured by Eastham town meeting action, he had more than enough to do. Gibbs was also acutely aware that how the Seashore handled its initial responsibilities in Eastham might well determine the future course of action in the other communities. He wanted the first beach development to be accomplished with consummate care and sensitivity.[83]

A second federal operation was also just around the corner. After many delays, the transfer of the state lands in Provincetown and Truro was imminent. The Seashore would fall heir to the developed state facilities at Herring Cove in Provincetown, but the master plan also anticipated a second beach facility in the spectacular dune area at Race Point. The state lands transfer was not without its headaches, however. Both at the Province Lands and at Pilgrim Spring State Park, the National Park Service had inherited a difficult problem of squatters on public land. The commonwealth was more than grateful to turn over these difficult public relations problems to the federal government.[84]

But the state lands transfer prospect did stimulate detailed discussion by the Advisory Commission on the formal establishment of the Seashore. A White House ceremony, marked by the execution of the state deeds by the governor of the commonwealth and the formal establishment of the Seashore by the secretary of the interior, seemed to make good sense. Regional Director Ronald F. Lee was asked to pursue this possibility with appropriate Washington officials. The answers were encouraging but inconclusive.[85]

By the end of 1963, the towns of Chatham, Orleans, Eastham, and Truro had adopted zoning bylaws that appeared to meet the secretary of the interior's standards. The required approval of the Massachusetts attorney general had been received for each of these actions. A special town meeting was planned for Wellfleet, but the prospects were still uncertain within that community. Influential town moderator and selectman Charles E. Frazier, Jr., had publicly expressed his indifference to the action. And despite more than eleven hundred acres of privately owned land subject to Seashore influence, the town of Provincetown had taken no action at all. In unprecedented action, the Advisory Commission voted to remind the Provincetown selectmen of this deficiency.[86]

An unexpected dividend of the zoning discussions had been the linkage with Blair Associates, the planning firm charged with the preparation of a master plan for Barnstable County. This prompted the Advisory Commission to consider the Seashore's relationship with other regional and master plan studies. Of particular concern was the transportation bottleneck represented by the two-lane mid-Cape highway, the only land approach to the Seashore. Following presentation by state highway officials at an Advisory Commission meeting, the National Park Service was urged to find ways to accelerate the highway improvement program.[87]

At a joint meeting held with the Minuteman National Historical Park Advisory Commission, the two bodies considered ways of joining forces to encourage the Department of the Interior to take an active interest in these external transportation problems, for a similar problem existed in Concord and Lexington. The joint meeting was useful in other respects as well, since it brought together individuals with significant interests in common. Among the members of the Minuteman Advisory Commission were Mrs. Katharine White, the daughter of Cape Cod senator Edward C. Stone, and Representative James DeNormandie, an early champion of the Cape Cod National Seashore in the Massachusetts legislature.[88]

There was other evidence also that the Cape Cod National Seashore Advisory Commission was winning respect nationally. The citizens committee for the proposed Fire Island National Seashore in New York came to visit and returned much impressed with the Cape Cod precedent. Priscilla Redfield Roe, the daughter of Woods Hole scientist Alfred C. Redfield, had just moved to Long Island and was becoming influential in the Suffolk County League of Women Voters. She had cut her teeth on conservation issues in the Sudbury River valley and the Cape Cod National Seashore.[89] Farther west, the Lansing (Michigan) *State Journal* had editorialized in favor of the proposed Sleeping Bear Dunes National Lakeshore, recommending that the so-called Cape Cod formula be considered for this national park legislation.[90]

By early 1964, a benchmark of sorts had been reached. The Commission had moved from the basement of the Nauset Life Boat Station to the second floor of the renovated garage. A new member had joined the Commission—Harold J. Conklin, who was replacing Truro representative John R. Dyer, Jr., forced to resign by the demands of his flourishing legal business. The National Park Service had undergone changes also. Its new director George F. Hartzog and his family would make an unscheduled visit to the Seashore during the summer.

Vice-Chairman Nickerson brought back to the Advisory Commission detailed accounts of his comparative evaluation of Cape Cod and Everglades National Park in Florida. His host on that occasion was Everglades superintendent Stanley C. Joseph, the man destined to replace Robert F. Gibbs as superintendent of the Seashore before the year was out.[91]

The tragic assassination of President John F. Kennedy in November of 1963 had placed the Cape Cod National Seashore Advisory Commission in a delicate situation. There was considerable national sentiment to rename the Seashore in his memory. Other suggestions included a set of Kennedy memorial archives within the Seashore, or as the president of the Cape Cod Chamber of Commerce wrote the Commission, a memorial walking trail enabling the public to recapture the essence of the popular photograph of President Kennedy enjoying the dunes of his favorite Cape Cod. The Advisory Commission made its own recommendation to the Seashore, however, endorsing Chief Naturalist Gilbert's imaginative idea of an underwater viewing trail at the Salt Pond Visitor Center, a project regrettably never implemented for lack of funds.[92]

In the meantime, land acquisition was proceeding at a steady pace. The state lands transfer deeds had been recorded even without the desired establishment ceremony. The National Park Service had won a major appropriation battle in Congress, receiving an additional $4 million in funds for land, including a $1 million transfer from Minuteman National Historical Park. A substantial decentralization of land acquisition authority had taken place enabling the Northeast Regional Office to approve land transactions up to a maximum of $200,000. However, the most significant land acquisition development had been congressional passage of the Land and Water Conservation Fund. This act earmarked certain revenues from offshore oil and gas leases and from the sale of surplus federal properties for a special fund for the acquisition and development of recreation lands at state and federal levels. For the first time, it would be possible to develop a long-range acquisition program for the Seashore.[93]

With the Seashore now in full swing, a host of operating problems had arisen. Winter storms had given the Seashore its first experience with snow removal. More than fifty unsightly buildings, remnants of the former Camp Wellfleet, had been removed under contract or with contract personnel. Despite some misgivings on the part of the Advisory Commission, the American Youth Hostel had been issued a permit to utilize the former Coast Guard station in Truro, known popularly as "Little America," as a seasonal headquarters for youth groups. Special facilities for the blind and the disabled were under active consideration, and Henry Beston's "Outermost House," now property of the Massachusetts Audubon Society, had been formally dedicated. The Seashore was clearly growing in popularity, with the new Race Point bathing area filled to capacity by noon on good days.[94]

A lower Cape organization, the Association for the Improvement of Medicine (AIM) on Cape Cod, had provided the Advisory Commission with its first political test. Despite a clear consensus favoring the improvement of medical resources, the Commission felt, and Interior concurred, that the proposed medical facility should be located outside the Seashore's boundaries.[95]

A policy problem identified earlier by Regional Director Ronald F. Lee continued to fester: Should facilities for public camping be developed as part of the Seashore? The Advisory Commission decided to address the issue in a series of meetings, inviting private campground owners, business concerns, and local and state officials to attend specific sessions to discuss matters of demand, supply, and policy. The outgrowth of the discussions was a reaffirmation of the earlier policy encouraging the continuation of private campground facilities within the Seashore. Developed campgrounds under federal auspices were to be only a last resort.[96]

Toward the end of the year, *Boston Globe* Cape Cod correspondent Frank Falacci complained in print that there had not been a decent fight for months on Cape Cod, recalling wistfully the earlier heated battles over the establishment of the Seashore. He described Wellfleet's Esther Wiles as "the last of the Cape's great fighters," and indeed she was. Mrs. Wiles continued to speak out forcefully against what she regarded as excessive land acquisition, served as the spokesman for many of the commercial interests within the Seashore, and took a dim view of the educational programs because of what she termed their potentially debilitating influence on youth and their alleged contribution to the disintegration of family authority.[97]

It was also clear that the fight was not completely gone from the Advisory Commission either. The Seashore's architectural plans for the Salt Pond Visitor Center produced a heated debate about architectural style. More than a few acid comments were made about "that beehive thing" destined to command the horizon overlooking scenic Salt Pond. Why not a nice, simple, Cape Cod cottage, Chatham's McNeece asked?[98]

By late 1964, the Seashore had fallen heir to one of the first four conservation centers established by the Office of Economic Opportunity under the Job Corps program. The Cape had been generally opposed to this social legislation, and the Seashore administration was hardly enthusiastic about another major enterprise on top of the responsibilities it already had. Nevertheless, by early 1965 a fifty-enrollee camp was operating, and a small work force of disadvantaged youths from the inner city was available to help with Seashore and local conservation projects. Once established, the Job Corps Conservation Center won surprising local support. Superintendent Gibbs expressed himself as "amazed and grateful" when the Job Corps was invited to submit a float in every one of the Fourth of July parades on the lower Cape.[99]

Seashore attendance continued to climb. Even in January, some thirty-nine thousand visitations were recorded within the Seashore. Additional ocean beach and parking facilities were clearly needed. Head of the Meadow Beach in Truro, former Camp Wellfleet, and Great Island in Wellfleet were obvious possibilities for future expansion. Of them all, Camp Wellfleet appeared to offer the most potential.

During the consideration of development alternatives, several issues came up. The sections of the Seashore Act assuring continued local shellfishing were put to the test at Great Island. After much debate, the Advisory Commission recommended that local shellfishermen be given special access privileges at Great Island for certain months of the year.[100]

The Great Island issue also brought into the open a matter that had been simmering on several fronts. Use of Cape Cod beaches by oversand vehicles had been traditional for many years, but obviously represented a conflict with not only the basic preservation mission of the Seashore but also the activities of other user interests. The situation was particularly acute on Nauset Beach, property technically within the boundaries of the Seashore but still under the control of the towns of Orleans and Chatham. Beach buggy interests were dissatisfied with local regulations and were threatening to petition the legislature for a state park on this site. The situation was compounded by the fact that the Seashore's oversand regulations were appreciably different from those of the local communities.

From the Advisory Commission's viewpoint, however, the oversand issue had to be set aside in favor of still another pressing problem. As Vice-Chairman Nickerson reported to the Commission, the new federal Recreation Advisory Council, established under the Land and Water Conservation Fund Act, had just issued Policy Circular 1. Along with other areas, the Cape Cod National Seashore had been categorized as a recreational area.

Commission concern over the possible dilution of the conservation emphasis of the Seashore led to repeated correspondence and numerous meetings at the Washington level. Although the formal designation was never changed, the Advisory Commission won agreement in principle from Director Hartzog that the legislative language would prevail. The agreement was fortified by special language in the fiscal year 1966 appropriations act inserted by Senator Leverett Saltonstall expressing the underlying conservation philosophy of the Seashore. As Vice-Chairman Nickerson observed: "This puts an end to it . . . for now."[101]

By early 1966, the formative stages of the Seashore were clearly over. To emphasize this fact, word was passed that the Seashore would be formally established. To the Advisory Commission, the event was welcome but somewhat anticlimactic. By the date of the establishment ceremony in May of 1966, substantial changes were either accomplished or underway.[102]

After more than thirty years of public service, Superintendent Gibbs had announced his retirement. So had Regional Director Ronald F. Lee. Chief Naturalist Vernon G. Gilbert was leaving for a special assignment in Africa. It was also rumored that Chairman Foster would be leaving shortly for a new assignment in Washington. The real viability of the Seashore would certainly be put to the test as the Advisory Commission began its third two-year term early in 1966 under substantially different management.[103]

THE MATURING YEARS

Robert Gibbs's replacement, Superintendent Stanley C. Joseph, described as a "mature and able administrator" by National Park Service director George F. Hartzog, Jr.,[104] pledged his best efforts during his first meeting with the Advisory Commission. "The most information is the best information," he observed.[105] His counterpart at the regional level, Regional Director Lemuel A. (Lon) Garrison was also an experienced, career Park Service employee. Joseph and Garrison had first worked together at Yosemite National Park in 1935.[106] There was some grumbling on the Commission's part about the frequency of personnel transfers. Cape Cod had clearly become a launching pad for Park Service professionals. As one member put it, perhaps the Advisory Commission "wears them out."[107]

But the Advisory Commission itself had undergone significant change. After much reflection and some hesitation, Secretary Udall had appointed Barnstable County representative Joshua A. Nickerson as the new chairman.[108] The Advisory Commission had responded by electing Interior representative Leo F. Diehl as its new vice-chairman. And as the year began, Truro made its third change on the Advisory Commission, nominating Selectman Stephen Perry to replace retiring Harold J. Conklin.

In Superintendent Joseph's words, the Seashore "had truly taken form" in 1965.[109] The increase in visitor use had been dramatic. The Seashore was being literally overwhelmed by visitation. There was an acute shortage of beach space, a growing traffic problem, accelerating use of the limited bicycle trails available, and much vandalism of Seashore property.[110] The new recreational craze—surfing—had accounted for five thousand visitor days alone during 1965, and there was a distinct need for special regulations both within the Seashore and on town beaches.[111] Perhaps the most graphic appraisal of visitation was supplied by Wellfleet's Esther Wiles, who reported one thousand six hundred cars passing her roadside jelly stand one hot summer day.[112]

Predictably, the development program of the Seashore occupied much of the attention of the Advisory Commission during 1966. High course tides had seriously damaged the existing facilities at Coast Guard Beach in Eastham. The Advisory Commission called in experts from the Corps of Engineers and the Woods Hole Oceanographic Institution for comment and advice. The recommendations were unanimous: work with nature not against it.

Another difficult issue involved the future of the nine-hole golf course at Highland Light in Truro. The property had been acquired by the National Park Service, and there was a question as to whether the golf course should remain. It was suggested that the town of Truro seriously consider operating the golf course as a municipal venture to provide an indirect means of obtaining reimbursements in lieu of taxes. With town meeting approval, Truro officials sought official sanction.[113]

A proposed second visitor center in the Province Lands revived the previous debate over architectural style. Even more significant than this "spider crab" on the dunes,[114] Commission members felt, was the question of how much development the Province Lands could stand. Among others, the Provincetown Conservation Commission was growing concerned about the threats to natural values posed by the Seashore's developed facilities.[115]

Unwittingly, the new superintendent advised the Commission that plans for a fifty-car parking facility at Camp Wellfleet were being readied for public bids.[116] The parking lot was to service not only beach users, but a proposed ocean tour road starting from Eastham and ending at Camp Wellfleet. The Advisory Commission's reaction was immediate. Land was not yet available for either the tour road or the entrance road to the proposed parking facility. There appeared to be little coordination between the land acquisition and development phases of the Seashore. The new superintendent was reminded firmly of the provisions of section 8(g) of the act, which required him to seek the advice of the Advisory Commission before proceeding with recreation facility development.

Outside of the procedural questions, it was also evident that some difference of opinion existed within the Advisory Commission on the most desirable type of development within Camp Wellfleet. Chairman Nickerson spoke eloquently of the need to protect the "Thoreau quality" of the high dune section of Camp Wellfleet. Other Commission members felt that the Camp Wellfleet development provided an ideal opportunity to enter into a package land exchange with the town of Wellfleet for scattered interior parcels. It was estimated that by using town lands for the entrance road, the Seashore could save approximately $25,000 in development costs.

As the discussion continued, it became evident that the Advisory Commission regarded Camp Wellfleet as the single highest priority development project for the Seashore. Superintendent Joseph promised to defer the parking facility in favor of a detailed development plan for Camp Wellfleet as a whole. The plan, the Advisory Commission felt, should also include a contingency provision for public campgrounds should the present policy permitting only private campgrounds become intolerable. The unfortunate result of not consulting with the Commission at the outset on priorities, Chatham's McNeece observed, was at least a year lost on needed facility development.

On the land acquisition front, price escalation had nearly eaten up all of the Seashore's authorization. Approximately $1.5 million remained for land acquisition. There was the possibility of an additional $450,000 through transfer, but beyond that, all of the funds would be gone with some eight thousand acres of Seashore lands still to be acquired. A closing of the lands office was imminent, and this raised serious questions about the future of George H. Thompson, an individual whom the Commission regarded as absolutely essential to the completion of the land acquisition program. In recognition of these circumstances, Congressman Hastings Keith had filed legislation increasing Cape Cod's authorization by an additional $12 million, but prompt action by Congress appeared unlikely because of difficulties faced by the Land and Water Conservation Fund nationwide.

The shortage of land acquisition funds had created other problems as well. As commercial permits came up for renewal, the Advisory Commission had no choice but to recommend approval. If no money was available for unimproved lands, the Seashore certainly could not acquire commercial properties. There was also the uncomfortable specter of land speculators and developers taking advantage of the Seashore's predicament. Holding the line on zoning provisions might not be possible if the Seashore was unable to meet a development threat with immediate condemnation.

Other land acquisition projects were not going well either. It was clear that the towns of Truro, Orleans, and Chatham would continue to hold their town beaches. In the case of Wellfleet, despite the urgent need for town-owned property for the Camp Wellfleet development, Commission member Esther Wiles had been appointed chairman of a special Wellfleet Lands Exchange Committee and was engaged in some hard bargaining with the Seashore. Her proposed exchange for lands along the ocean front was in direct contradiction to the policies of the Seashore and could not be accommodated.[117]

One other land exchange was proceeding, however. The town of Eastham had proposed an exchange enabling the relocation of the Nauset Regional High School on lands within the Seashore's boundaries. This seventy-acre land exchange could well eliminate many of the remaining inholdings under the control of the town of Eastham.

With its finely tuned antennae, the Advisory Commission had become alert to another coming problem—that of offshore oil and gas development—"a pretty big fight" in the words of Chatham selectman and Commission secretary Robert McNeece.[118] Chatham's interest derived from a small fishing fleet and a particularly active lobster fishery off the outer beach of Cape Cod. The *Torrey Canyon* sinking had been given widespread coverage by the world press, and Cape Cod itself was experiencing growing problems with oil slicks on the beaches from passing coastal tankers. Concerned over the seeming indifference of the Department of the Interior and Congress to the implications of offshore oil development, the Advisory Commission named McNeece a committee of one to press for protection of the Seashore at the state level at least. Within a year, McNeece could report the establishment of an official Cape Cod Ocean Sanctuary restricting all such mineral development activity within the state's three-mile limit.[119]

By the close of 1967, a previous issue reared its head again. In a letter to Chairman Nickerson,[120] Cape Cod Chamber of Commerce executive director Norman Cook pointed out the precedent of payments-in-lieu-of-taxes presented by the new federal legislation authorizing the Redwood National Park. Why should not Cape Cod seek similar provisions, he asked? Wellfleet representative Esther Wiles agreed. At her request, the Advisory Commission voted to seek a system of payments for federal properties leased by the Seashore or under long-term private occupancy. Faced with a negative response from Interior officials, Truro representative Stephen Perry was given the assignment of seeking appropriate state legislation. After a hard battle in the state legislature, the so-called Aylmer Bill was enacted into law, and the Cape Cod communities won the right to issue tax bills to property owners enjoying use and occupancy privileges within the Seashore.[121]

By 1968 the Cape Cod National Seashore was facing an accumulation of crises. The land acquisition office was closed out in mid-April with the additional authorizing legislation seemingly stalled at the congressional level. There were significant delays in condemnation settlements and severe price escalation in the interim. Threats of development were reported in several communities. Emergency funds would be needed if the new legislation was not passed promptly. Negotiations with Wellfleet had broken down completely.[122] The impasse on land exchange had forced the Seashore to design around the town lands for the prospective beach development at Camp Wellfleet.

The Commission's initial regional planning venture in conjunction with the new Cape Cod Planning and Economic Development Commission had been received with considerable skepticism. Despite the obvious need for cooperative action in such matters as transportation, town officials were wary of regional entanglements. Other Commission members like Wellfleet's Esther Wiles regarded regional planning as the first step toward communism and slavery.[123]

An intransigent Provincetown Airport Commission was proposing a series of development projects within the Seashore, seemingly unsympathetic to conservation considerations. And the matter of oversand vehicle regulations was of continuing concern. On top of all of these difficulties, the Seashore was faced with its first significant budgetary cutback. The most tangible manifestation was the closing of the Salt Pond Visitor Center two days a week outside of the summer season.

Yet there were also some high points in what was otherwise a discouraging year. The Scientific Advisory Committee, first proposed in conjunction with the erosion problems at Coast Guard Beach in Eastham,[124] had started to flourish under Josiah Child's direction. Twelve distinguished Cape Cod scientists had agreed to serve on the Advisory Committee. Their first areas of concentration had been Great Island, which the committee felt should remain in a natural condition, the dunes section of Provincetown and Truro, and the deteriorating condition of Pilgrim Lake. A particularly active member of the Scientific Advisory Committee was Dr. Norton H. Nickerson of Tufts University, a resident of the town of Dennis. When Josiah H. Child announced his retirement to Florida and his unavailability for further Advisory Commission service, it was only appropriate that Dr. Nickerson take his place as the commonwealth's citizen representative.[125]

In October of 1968 the Seashore was visited by a delegation from the Indiana Dunes National Lakeshore Advisory Commission. Five of its members, accompanied by Park Service representatives, arrived by private jet for a tour of the Seashore and a visit with the Cape Cod Advisory Commission. As Regional Director Garrison later reported to the director, the occasion revealed many areas of common concern and a "happy note of shared pride."[126] Outside of the effectiveness of the Advisory Commission itself, the Indiana Dunes visitors found two elements of the Cape Cod Seashore particularly outstanding: its central architectural theme and the generally high quality of its development.

An updating of the master plan had been scheduled and an outside review committee appointed.[127] The process now involved thirteen steps in all, six of which must be conducted in the field. The first step was a policy document forwarded to the Commission for detailed review. The Advisory Commission approved its contents and noted with particular interest the prospective designation of five environmental study areas within the Seashore and the adoption of the University of California's National Environmental Education (NEED) program of interpretive education.

The Advisory Commission was saddened by the news of the death of Arthur Finlay, Orlean's original representative. It was equally disturbed to learn of the pending retirement of Superintendent Joseph and the arrival of the third superintendent for Cape Cod in three years' time. The new superintendent was to be Leslie P. Arnberger, another career Park Service employee with many superintendencies already to his credit. The Advisory

Commission was comforted by former superintendent Robert Gibbs's evaluation of Arnberger: "They couldn't have found a better choice."[128]

By 1969 Superintendent Arnberger could report most of the remaining development projects well under way. The new Marconi Beach at Camp Wellfleet was scheduled to open on July 3. The new Province Lands Visitor Center had been dedicated on May 25 with an impressive group of public and local officials in attendance, including Governor Francis W. Sargent, Under Secretary of the Interior Russell Train, and former senator Leverett Saltonstall. What the Advisory Commission had derisively termed the "Chinese pagoda" had been pressed into useful service almost immediately.[129] The Visitor Center's outdoor amphitheater was ideal for official ceremonies, but also for community events ranging from the Provincetown Symphony Orchestra to rock concerts.

The NEED program was clearly on its way to becoming an outstanding success. A group of fifth graders from the Andrew School in South Boston were its first visitors. By the end of the second year, all thirty-six weeks were committed to school groups. The pending closing of the Job Corps Conservation Center raised an obvious possibility—using some of the buildings for an expanded NEED program, particularly by the Falmouth school system which was making such extensive use of the field education program. The Advisory Commission remained highly supportive of these urban education programs[130]—at one point inviting Old Dominion Foundation executive Ernest Brooks, an old friend of Cape Cod, to a Commission meeting to discuss the need for additional resources. As Chairman Nickerson observed, it is important to teach people that water is not produced by faucets nor light by switches. Superintendent Arnberger pledged his personal efforts to obtain a suitable program for the former Job Corps Conservation Center. By year's end, he had persuaded the Nauset Regional School District to consider using one of the buildings as a clinical teaching center for disabled and disadvantaged children.[131]

By 1970 good news had reached the Advisory Commission. At long last Congress was proceeding with the land acquisition authorization legislation. Malcolm Dickinson, chairman of the Orleans Conservation Commission, and Warrenton Williams, chairman of the Cape Cod Planning and Economic Development Commission, reported on the House and Senate hearings. It was now certain that the Seashore's authorization would be increased to $33.5 million, more than double the original authorization.

Dickinson and Williams also reported on another matter considered by the congressional committees—the possible extension of the ten-year term of the Advisory Commission. While this matter was not included in the ultimate authorization legislation, the congressional committees were reported to be fully aware of the extensive contributions of the Advisory

Commission. As Williams expressed it, Congress was prepared to make the change "when the time is due" so that the Advisory Commission is not "left out in the cold."[132] The decisive factor on the authorizing legislation was testimony by the National Park Service that extension was not currently needed since the Advisory Commission still had some six years of its statutory life remaining.

Superintendent Arnberger informed the Advisory Commission that the Service would move promptly for appropriations within the new authorization ceiling. He was as good as his word, for the fiscal year 1971 appropriation included $8.8 million for Cape Cod, of which $1 million was released immediately by the Bureau of the Budget.

Arnberger reported that Richard Schwartz, a land acquisition officer with recent experience at Fire Island, would head up the new Seashore acquisition program together with a staff of seven. It was estimated that some two thousand tracts remained to be acquired. Exclusive of the 1,300 acres of improved property remaining within the Seashore and the 2,700 acres of town-owned land, the job ahead approximated 3,500 acres.[133]

The second-phase land acquisition program, outside of purchase of the remaining unimproved property, would have one significant feature, the Advisory Commission determined: acquisition of all properties developed after the cutoff date of September 1, 1959, provided in the Seashore enabling act. In terms of commercial property, the policy would be one of case-by-case analysis. No purchase of these commercial properties would occur unless they were offensive in some fashion or inconsistent with the master plan objectives. The Advisory Commission contemplated a continued measure of commercial operations within the Seashore and even an expansion of those operations compatible with Seashore interests.

The Advisory Commission expressed considerable concern that an approved master plan was still not in existence after seven years' time. The draft master plan finally brought before the Commission for review and approval limited future development to bicycle trail expansion and picnic area development. Its primary emphasis rested heavily on interpretation and education. A new visitor center was proposed for Griffin Island in Wellfleet to educate visitors to the features of the bay side of the Cape. The infamous tour road would be eliminated. Under new procedures, the draft master plan was subjected to public hearing with nearly 250 local citizens in attendance.

It was also reported that the National Park Service had contracted for an update of the earlier economic evaluation. The firm of Philip B. Herr and Associates had prepared a report on the economic effects of the Seashore during the period 1961–68.[134] The indicators were all positive. Predictably, private land values had increased 106 percent since 1961, with tourist industry employment and wages up approximately 50 percent. Assessed valuations within the communities had risen 11 percent per year. Property taxes had dropped 16.9 percent. The Herr report could attribute $10 million in direct sales to Seashore visitors. Three million dollars had been spent for Park Service facility development and $16 million for land acquisition. Payroll expenditures for the Seashore were at a level of $300,000 to $500,000 annually.

The economic planners concluded that the chief economic impact of the Seashore lay in the increase in private land values on the lower Cape. The chief overall impact of the Seashore was environmental preservation. Although much of the profitable business increase would probably have occurred without establishment of the Seashore, the land value, the motel and restaurant business, and the permanent population increases were definitely the results of the Seashore.

By 1971 the Advisory Commission had settled into an efficient manner of operation. With the volume of business somewhat reduced, the Commission had begun meeting only bimonthly. Its equanimity had survived sensitive issues such as community rock concerts, the institution of beach facility charges, problems of low-flying airplanes, and even the first female life guard clad in a flame orange swimsuit. But the potentially disturbing problem of nude bathing had begun to emerge, a long-standing practice on Cape Cod but one whose devotees were growing in number.

In the meantime, the Advisory Commission had suffered the loss of another longtime member—Interior representative Leo E. Diehl, the vice-chairman of the Commission. With a change of administration in Washington, Diehl had been replaced by his Cape Cod friend and neighbor, Chester A. Robinson, Jr.[135] Selectman Linnell Studley had become the new representative from Orleans. As Superintendent Arnberger presented a replacement gavel to Chairman Nickerson to "keep his unruly members in line,"[136] the Commission was prompted to ask of itself whether all terms should expire at once as was the present practice and whether a deliberate effort should not be made to obtain a complement of younger members. This was particularly timely because a new regional director, Chester Brooks, had been appointed and the whole matter of advisory committees was under intensive examination at the federal level.

By mid-1971, significant accomplishments were reported on the land acquisition front with two large shorefront properties, the Ball Trust holdings and the Mitre tract, now safely in Seashore hands. Although negotiations with the post-1959 property owners were difficult, the blow had been softened by a policy decision to offer owners a twenty-five-year occupancy privilege to ease their transition problems. In addition, the provisions of the Uniform Relocation Act of 1970 made it mandatory for the National

Park Service to find suitable replacement accommodations for individuals displaced by any Seashore acquisitions.[137]

The conflict with oversand vehicle user interests continued to smolder, culminating in the submission of state park legislation for Nauset Beach by the Massachusetts Beach Buggy Association (MBBA).[138] This action prompted intensive discussions by the selectmen of Orleans and Chatham. Both towns authorized their selectmen to explore with the National Park Service a transfer of these beach properties to the Seashore.[139]

At the request of the Advisory Commission, the National Park Service developed a draft set of management objectives that emphasized the preservation of the semiwild character of Nauset Beach if the land should become federal property. The state legislature, however, aborted the issue by establishing a special Commission on Ocean Beaches in lieu of action on the MBBA proposal. The issue, however, did result in the execution of a cooperative agreement on use and management between the towns and the National Seashore. This agreement, plus the new state regulations issued by the Division of Marine and Recreational Vehicles and the Department of Public Health, would ensure a greater level of compatibility between federal, state, and local beach administrations.

Prompted by these special concerns, the Advisory Commission believed it was time to gather together town officials to discuss generally the nature of cooperative activities. Among obvious matters of concern were police and fire protection, but repeated outbreaks of gypsy moths, mosquitoes, and other nuisance insects dictated an examination of these protection needs as well.

From Provincetown's official representatives came an expression of still another concern—that of future water supply. Pressures on available sources had prompted Provincetown to seek a special permit from the Seashore to explore for additional wells in the Truro portion of the lower Cape. The Advisory Commission encouraged a cooperative water resources survey by the U.S. Geological Survey to assist the towns in meeting their own needs, pointing out that the Seashore itself was dependent upon town wells for its basic water supply. From preliminary reports, it was obvious that the water supply situation was one that needed to be faced on a countywide basis. New member Norton H. Nickerson, also active at the state level in conservation commission and natural resource programs, offered to serve as liaison for the Advisory Commission in this regard.

By early 1972 a collateral issue had surfaced within the Advisory Commission. Senator Edward M. Kennedy had introduced legislation calling for a federal Islands Trust encompassing Nantucket, Martha's Vineyard, and the Elizabeth Islands. Earlier versions of legislation had suggested that the Cape Cod National Seashore might be expanded to include these properties, and Chairman Nickerson had been instructed to stay abreast of all such developments. Both Nickerson and Superintendent Arnberger were invited to Nantucket and Martha's Vineyard to discuss the Cape Cod experience. Nickerson refuted a Vineyard newspaper characterization of himself as a reformed opponent of the Seashore, stating that his position had changed only because he had recognized the finality of the Seashore's existence and the need to make it as effective as possible.[140]

Of more direct concern to the Advisory Commission's existence was the passage of the Federal Advisory Committee Act by the Ninety-second Congress.[141] This followed on the heels of Executive Order 11671, which placed all advisory commissions under Bureau of the Budget supervision.[142] By the act's provisions, all nonlegislative commissions would terminate on January 5, 1973, and all legislatively established commissions would expire at the end of their authorized period. Previous legislative language notwithstanding, even the statutory advisory commissions would become subject to Interior and Bureau of the Budget management under a set of uniform guidelines. Among their provisions were some of those deemed so onerous at the time of the Commission's establishment. The statute also contained a requirement that all meetings of advisory commissions would be fully open to the public and the press.

The Cape Cod Advisory Commission was informed that it could not meet officially until properly chartered by the Department of the Interior.[143] Since Interior continued to have distinct reservations about the continuation of individual park advisory committees,[144] it was likely that the Cape Cod Commission would be phased out entirely at the end of its authorized period and its functions assumed by an advisory committee attached to the regional office. The northeast region of the National Park Service had been favored with the establishment of the first regional advisory committee, but the Cape Cod Commission did not believe its interests would be represented adequately despite its former chairman, Charles H. W. Foster, being a member of this body.

With these straws in the wind, the Advisory Commission moved on to its one hundredth official meeting on September 8, 1972. Former members Josiah H. Child and John Carleton arranged to be present. Commonwealth representative Arthur W. Brownell flew in by helicopter with former chairman Foster, the new cabinet secretary of environmental affairs for the commonwealth of Massachusetts. A particularly poignant part of the ceremony was the retirement of Robert McNeece as a Commission member and its secretary. Letters from Secretary of the Interior Rogers C. Morton and Governor Francis W. Sargent and a scroll containing the signatures of every member of the Advisory Commission were presented to McNeece, the member who had missed but two meetings of the Advisory Commission in its entire ten-year history.

After a brief moment of nostalgia, the Advisory Commission proceeded to elect Orleans representative Linnell Studley as its new secretary and welcomed Selectman David F. Ryder as Chatham's new representative.[145] As an unscheduled anniversary issue, the Advisory Commission turned back the pages of history for a spirited discussion of camping. It was clear that attitudes had not changed appreciably in ten years' time.

THE COMING OF AGE

In Superintendent Arnberger's annual report to the Commission,[146] it was plain that the Seashore had come of age. Statistics disclosed almost five million visits to the Seashore in 1972. The interpretive program had grown from seven walks a week in 1963 to fifty walks a week ten years later. Not only was the former Nauset Coast Guard Station being employed to capacity by the NEED program, but "Little America" was also being used in the winter for similar purposes. And the Falmouth school system had expressed interest in redeveloping several of the Mitre buildings for its own version of the environmental education program.

Matters relating to oversand vehicles had settled down to an uneasy truce. The Seashore's regulations had been sustained by the U.S. Magistrate, and the efforts to explain them persuasively to beach buggy enthusiasts had been relatively successful. Planning was under way for the celebration of the seventieth anniversary of the Marconi Station, site of the first transoceanic wireless transmission, with interest in the event exhibited literally around the world. Land acquisition was progressing well, with less than a thousand of the twenty-six thousand acres left for federal purchase. It was understood that these acquisitions would take time because many small ownerships were involved, such as those at Nauset Marsh with uncertain or unknown title. The unexpected had also continued to occur within the Seashore. There had reappeared on Peaked Hill Bars off Provincetown the HMS *Somerset*, the British frigate sunk nearly two hundred years ago with all hands on board. It had become a mecca for the skin diver and the summer visitor.

The primary business at hand was the future development of Coast Guard Beach in Eastham. Despite a 1970 congressional authorization for a beach erosion study by the Corps of Engineers, the Seashore was reconciled to a method of development that would deliberately anticipate continuing erosion of the shoreline. Park Service planners had formulated a series of development options, including shuttle bus service to the beach from the main parking area at the Salt Pond Visitor Center. Eastham officials were skeptical of the viability of this proposal, wondering how a family destined for a day on the beach could possibly carry all its equipment on a shuttle bus. The most plausible alternative appeared to be a series of overflow parking lots adjacent to Coast Guard Beach which would lead the visitor ultimately to the ample facilities at Camp Wellfleet as each parking facility was filled to capacity. Particular features of the development scheme were the ocean viewing areas for casual visitors and the bicycle paths in Eastham connecting the Salt Pond Visitor Center with the ocean shoreline. Superintendent Arnberger, however, warned the Advisory Commission that development money was tight because of growing fiscal stringencies. Only reprogrammed funds would be available for development.

At its meeting in September of 1973,[147] the Advisory Commission received testimony from Adrian Murphy of Truro, the chief proponent of the AIM proposal in earlier years and now a spokesman for the Truro Neighborhood Association. Murphy expressed the Association's growing concern over the so-called free beach, the area at Brush Hollow in Truro used historically by nude sun bathers. "This is a matter of growing concern," Murphy said, "which can't just be swept under the rug."

Commission member Dr. Norton Nickerson spoke of his own experiences during the summer at nearby Ballston Beach, and Wellfleet's Esther Wiles expressed her concern that the nude bathing would spread into Wellfleet. Truro's Stephen Perry blamed a Boston Channel 7 television announcement for having swelled the normal fifty-person population at Brush Hollow to over two hundred.

Provincetown member Nathan Malchman, long accustomed to such problems in his own community, stated that there was no way to legislate morality. He was supported in this observation by Superintendent Arnberger who advised the Commission that nudity per se was not against the law. The only recourse the Seashore had was to charge such people with disorderly conduct. The Advisory Commission discussions, however, did serve a useful purpose. All hands agreed that the Seashore should be prepared to face this issue squarely in advance of the 1974 season.

But by 1974, change was in the wind again. Superintendent Arnberger was to be transferred to Yosemite National Park, a unit of the National Park System urgently needing his particular skill and diplomacy.[148] His successor was to be Lawrence C. Hadley, an individual who had first come to know Cape Cod during his service in the director's office in Washington.[149] Planning was also under way to close down the second phase of the land acquisition program. Less than five hundred tracts remained to be acquired, and this job could be carried out by the Seashore's own staff.

Arnberger's departure provided an opportunity for the Advisory Commission to examine itself critically.[150] Was it still needed? asked Eastham representative Chase. What difference does it really make, Wellfleet's Esther Wiles responded tartly, since the National Park Service does what it pleases anyway. However, other Commission members expressed a decided opinion to the contrary. The Advisory Commission provides good citizen input, they said. They were supported by statements from the new regional director Jerry Wagers and his deputy regional director David Richie who told the Commission that it now enjoyed an enviable reputation as a prototype for other advisory bodies throughout the National Park System.

In a subsequent memorandum to former superintendent Arnberger,[151] Hadley confessed to a mixture of great interest and some apprehension at his first meeting with the Advisory Commission. In an aside, he offered the opinion that Wellfleet continued to appoint Mrs. Wiles as its representative just to make the superintendent's life more interesting.

The Advisory Commission was also to face a still-intransigent Provincetown Airport Commission which had insisted upon its runway expansion program despite the direct opposition of the Seashore.[152] The Scientific Advisory Committee advised the Commission that the airport expansion would create significant ecological damage. There was ample data to back up these assumptions, because the earlier scientific advisory effort had ripened into an official cooperative research unit with a full set of laboratory facilities at the former Mitre site in Truro. Under the direction of University of Massachusetts scientists Paul Godfrey and Alan Neidoroda, a team of summer student assistants was conducting research studies on such matters as beach and near-shore geophysical processes and the effects of groundwater withdrawals on marsh and wetland resources. Of particularly high priority were the research investigations of oversand vehicle usage. The preliminary results from the first year of the two-year study had persuaded the Seashore not to increase the number of beach buggy permits allowed on its ocean beaches.[153]

But as predicted, the nude bathing issue was to become the highlight of the Advisory Commission's 1974 sessions. Two open meetings were held. At the first session,[154] the Truro Neighborhood Association repeated its assertion that the free beach constituted an attractive nuisance. Local spokesman Stephen Williams, however, disagreed. He urged the Seashore to deal with the situation "in a creative manner." Having heard the public testimony, the Advisory Commission requested a statement of alternatives from the National Park Service.

At the second open meeting,[155] Superintendent Hadley advised the Commission that overuse of the area and not nudity was the primary issue. The options at hand were relatively simple: either accommodate the use appropriately or prohibit it entirely. The first course of action would lead to possible conflicts with other uses, as well as required changes in the master plan and current Seashore regulations. The second option would almost certainly lead to problems of regulation and enforcement and, quite likely, legal suits on civil liberties issues.

After much discussion, the Advisory Commission concluded that prohibition of nude bathing was the only practical recourse. In a comprehensive policy statement addressing the issue, the Commission recommended that the National Park Service proceed with the drafting of such a regulation to be applied "for at least the 1975 season."[156]

With legal assistance from the regional and Washington Solicitor's offices, the National Park Service responded with a draft regulation and the required environmental assessment of management alternatives.[157] The stage was now set for an almost certain confrontation during 1975.

But the Advisory Commission was facing a confrontation problem in reverse, for its long-standing member and critic, Esther Wiles of Wellfleet, had not been recommended for reappointment by the Wellfleet selectmen.[158] Her successor would be John Whorf, a retired businessman and director (with Joshua Nickerson) of the Lumber Mutual Fire Insurance Board.[159] Mrs. Wiles's outspoken comments and independent thoughts would be genuinely missed.

In transmitting the official resolution of the Advisory Commission,[160] Superintendent Hadley spoke of her dedication and perseverance, and the "Yankee understanding of self government" that had been so much of an ingredient in the acceptance and maturing of the Seashore. But as the *Province-town Advocate* later reported, Mrs. Wiles criticized the Seashore even on her way out.[161] She was not interested in serving on the Advisory Commission this year anyway, she said, because it was nothing but a "yes" commission.

With the Advisory Commission termination date just around the corner, it was time to concentrate on the matter of its future. Letters were sent to each appointing authority soliciting comments about the worth of the Advisory Commission and the desirability of its continuation. There was an immediate outpouring of praise for the Commission.[162] Its primary value appeared to be the liaison function it performed between the Seashore and the communities. As the Eastham selectmen put it, without the Advisory Commission the towns would have to deal with the Seashore "at arms' length."

Fortified by this ammunition, the Advisory Commission invited National Park Service associate director Robert Landau to attend the May meeting of the Commission.[163] Landau described the Cape Cod Commission as one of three excellent commissions throughout the National Park System, but spoke candidly of his own problem of how to handle the extension of one good commission without setting a precedent for all the others. He stated that the National Park Service would oppose any statutory period of extension. Chairman Nickerson and others responded with their concern about the Advisory Commission becoming a creature of the secretary of the interior and not of the Congress.

Yet at this same meeting of the Advisory Commission, Landau observed the firm backing of the Seashore by the Commission against any expansion of the Provincetown Airport runways, a position that led Commander McKellar of the Airport Commission to grumble that there were entirely too many environmentalists and conservationists today. Firsthand evidence of the Commission in action led Landau to advise Superintendent Hadley subsequently that he and new National Park Service director Gary Everhardt were in strong support of the administrative continuation of the Cape Cod National Seashore Advisory Commission.[164] With this encouragement Chairman Nickerson could report to the September Commission meeting that Senator Edward M. Kennedy was prepared to join Congressman Gerry Studds in filing legislation extending the statutory life of the Advisory Commission.[165]

But the big issue in 1975, as anticipated, involved nude bathing. On May 23, 1975, a suit was filed by the American Civil Liberties Union protesting the new Seashore regulations. Judge Frank Freedman visited the Brush Hollow site personally on June 23 and conducted a preliminary hearing in Boston on June 25. His decision was in support of the National Park Service. Although the federal district court ruling was appealed by the twelve plaintiffs in the suit, the new regulation would stand and would have to be enforced.[166]

By this time, the issue had been reported widely in the national press. Civil liberties proponents scheduled an organized confrontation for August 23, 1975. With the full support of the Advisory Commission, Superintendent Hadley instructed his rangers to be forceful but polite and courteous in enforcing the new regulation. The matter would be a difficult one to handle at best, for during the previous summer thirty-one separate nude bathing locations had been identified and a total of 611 warnings issued.

The morning of August 23, 1975, dawned bright and clear. The occasion was ostensibly a clothing-optional beach party sponsored by the Free the Beach Committee to which the public was invited. Network television and national newspaper reporters were on hand to record the celebrated confrontation. The approaches to Brush Hollow Beach were clogged with more than twenty-six hundred visitors. An additional twelve hundred onlookers were on hand. Acting upon instructions from Superintendent Hadley and his superiors, Park Service rangers merely directed traffic flow and maintained basic law and order. No warnings or citations were issued for violation of the nude bathing regulation. As Wellfleet representative

John Whorf later reported to the Advisory Commission, Seashore personnel exercised "admirable judgment."[167] By the next day the Seashore was back in normal operation, including the enforcement of all beach-use regulations.

THE SECOND DECADE

As the Advisory Commission assembled at Seashore headquarters in Wellfleet for its first meeting of 1976,[168] the topic uppermost in mind was the future of the Commission itself. Would it be rechartered as Robert Landau had indicated – "within weeks or days" of the expiration of its statutory termination date?[169] Commenting on the situation, *The Cape Codder* had said it well: the Advisory Commission must be prepared to "swap independence for life."[170] But there was still life *and* independence in the Commission. It had insisted that the Advisory Commission be chartered as a commission, not as a committee, and it had reinserted into the draft charter the excised statutory language requiring the National Park Service to seek the advice of the Advisory Commission before establishing a public-use area. By the April meeting of the Commission,[171] it was reported that the secretary of the interior had signed the revised charter and had cleared it for publication in the Federal Register. It was now time to turn to other matters.

Superintendent Hadley shared with the Commission the special dilemma he faced with respect to "improved properties" – those dwellings within the Seashore, constructed prior to the 1959 cutoff date, which were protected from condemnation by town zoning provisions.[172] Technically speaking, if a single board of the old structure was incorporated into the new building, a case could be presented for continuation of the Certificate of Suspension of Condemnation. Joshua Nickerson argued that Congress's intent was to retain a number of single-family, private dwellings, though not necessarily the same ones. Taken to its extreme, David Ryder observed, this could lead to homes only for the affluent. Superintendent Hadley felt a substantial measure of obligation, for the wrong advice on his part could subject a private dwelling to condemnation. It was high time, the Commission agreed, to develop revised guidelines for improved properties and to have the secretary of the interior suggest that the towns modify their zoning bylaws accordingly. The situation was particularly acute because the Seashore had run out of acquisition funds. The remainder of the $33.5 million authorized would be required for pending condemnation settlements.

By the April meeting, another matter of keen interest to Cape Codders had made its way to the Commission agenda – the proposed leasing for oil drilling on George's Bank, a mere hundred miles from the Seashore's famed ocean beaches. Roger Babb, the secretary of the interior's field representative in New England, and Frank Basile, the manager of Interior's Outer Continental Shelf Office, joined the Commission for a full and frank discussion of the issue. Babb lamented the present state of affairs, noting that he had spent an appreciable amount of his time "trying to counteract an unrealistic hostility based on misinformation and misinterpretation."[173] Massachusetts member Barbara Mayo was not impressed. Given the likely low yield of oil and gas from Georges Bank and the possible onshore effects from pipelines and spills, it seemed hardly worth the risks, she observed. Fellow biologist Norton Nickerson was in basic agreement. There is a world of difference between conditions encountered in the Gulf of Mexico, the source of Interior's optimism, and the more hostile environment of the North Atlantic, he commented. Chatham's David Ryder, a commercial lobsterman by trade, expressed skepticism over the assertion that the oil pipelines would remain safely buried in the shifting bottoms of Georges Bank. Before the meeting closed—the last under the Commission's statutory mandate—Chairman Nickerson expressed his personal thanks to the members present, and to the press, for their many contributions during a decade of activity.

It was not until October that the Advisory Commission could be reassembled again. The process of administrative reinstatement seemed interminable. As the presiding officer for the Department of the Interior, Superintendent Hadley read the new charter signed by Interior secretary Thomas S. Kleppe and introduced the new members of the Commission: Dexter M. Keezer of Truro; Clifford H. White, the new Massachusetts representative nominated by Governor Michael Dukakis to replace Norton Nickerson; and Dr. George M. Woodwell, head of the Ecosystems Center of the Marine Biological Laboratory in Woods Hole and the secretary of the interior's new designee. For one more term only Barnstable County representative Joshua A. Nickerson agreed to serve as the Commission's chairman. As Superintendent Hadley presented him the old gavel and block in recognition of past service, Nickerson observed that the Commission had been a very important experience in his life. Through it he had come to realize "how very, very wise it was to do exactly what has been accomplished by the Seashore," citing its primary contributions as making the Seashore a "viable part of the Cape Cod community" and promoting its "functioning along parallel lines with community interests." He felt himself "amply repaid" for the effort he had put into the Commission.[174]

But there was important work to be done and only two years of life under the Commission's new administrative charter. Chairman Nickerson lost little time in focusing attention on two management issues of growing concern to the Seashore. One was the matter of revised guidelines governing improved properties now exempt from condemnation; the other was the increased vandalism and trespass reported in the pond areas of Wellfleet and Truro where a mix of private and public jurisdictions made enforcement uncertain and difficult.[175] After a brief discussion of the principal issues, the Commission agreed to appoint special subcommittees to explore the matters in more depth. Dr. Barbara Mayo was asked to chair the subcommittee on ponds management, and fellow Provincetown resident Nathan Malchman, the newly reelected vice-chairman, agreed to head the subcommittee on improved property. By the December meeting of the Commission, it was clear that both subcommittees were hard at work and that their subject matter would be sensitive indeed. Like so many of the issues facing the Seashore, only the tip of the iceberg was immediately visible.

By its meeting in February of 1977, the Advisory Commission learned that the matter of modifications to "improved properties" had spilled over into the legislative arena. Congressman Gerry Studds and Senator Edward Brooke had filed bills, as *The Cape Codder* put it, to "correct abuses in the Park."[176] At Chairman Nickerson's suggestion, a formal statement was prepared for public release correcting the negative implications and detailing the Commission's consideration of the matter over the years. The new language allowing only forty-five days for action by the Advisory Commission seemed hardly realistic to most members given the cumbersome Federal Advisory Committee Act procedures.[177] In general, members could recall few instances of reluctance on the Commission's part to render advice promptly when it was sought.

Meanwhile, a new issue had reared its head—use of the former Province Lands at the northern tip of the Seashore. A local committee calling itself Save the Dunes addressed the April meeting on the subject of noise pollution from the Provincetown Airport. Its spokesman, Charles Schmid, also recommended creation of a wildlife sanctuary to protect endemic species from the pressure of off-road vehicles, hunting, and recreation. Save the Dunes was supported by John W. Grandy, executive vice-president of Defenders of Wildlife in Washington, who observed: "Because of the isolation and fragility of the dune ecosystem and because of the development over much of the lower Cape, the dunes and wildlife of the Cape Cod National Seashore more closely resemble an island than a continental ecosystem."[178] The Commission's response was to appoint a special subcommittee under the chairmanship of Clifford White, an ardent saltwater sportsman and beach buggy enthusiast, to look into the matter. By the June meeting of the Commission, White could report that the petitioners had a point. As many as seventy-eight sight-seeing flights were in the air on a given summer day. The Advisory Commission urged the superintendent to consult with the airport manager and try to arrange a reduction in the number of flights.

Business was brisk at the Seashore during the 1977 season, Superinten-dent Hadley reported, with visitation up a healthy 21 percent.[179] A new user group concerned with flying hang gliders had been accommodated thanks to cooperative action by the towns of Truro and Wellfleet.[180] Hadley also reported continuing interest in the Seashore from other park jurisdic-tions. Officials from Canada and from Trinidad-Tobago had visited during the summer. Dexter Keezer, the Commission's delegate to the new North Atlantic Region Advisory Committee of the National Park Service, added a footnote to Hadley's report. Keezer had found the regional committee to be an impressive group and was struck by the commonality of problems faced by park jurisdictions throughout the region. Yet, the Regional Advisory Committee, in his opinion, could never replace the Seashore's own Advis-ory Commission as a forum for free discussion and a way of "keeping open lines of neighborly communication between the Seashore Administration and the six towns."[181]

By the August meeting of the Commission, the Malchman subcommittee was prepared to offer its recommendations on the controversial matter of "improved properties." In a thorough report, the subcommittee traced the history of previous landmark cases, basing its review on the documenta-tion in the Seashore's files but also on a full day's field trip to visit thirty-six of the properties in question.[182] Seven specific recommendations were set before the Commission, the most urgent of which was the establishment of a reserve fund of at least $5 million to enable the Seashore to enforce the zoning provisions with condemnation proceedings if necessary. No legis-lation would be required if the secretary of the interior and the towns com-bined to promulgate and enact revised zoning standards and ordinances and the budgetary committees of the Congress proved cooperative.

The Malchman subcommittee report was also the primary topic of dis-cussion at the October meeting. Most controversial was the assertion that "any attempt to legislate so-called traditional architecture would be an ex-ercise in futility, not to mention the question of individual rights."[183] Dex-ter Keezer disagreed vehemently. Why not then a Moorish castle design or a far-out assorted angle modern design, he asked? Messrs. Ryder and Studley spoke of Chatham's and Orleans's experience in creating local ad-visory commissions to review commercial architecture.[184] By a vote of six to two, the Malchman subcommittee report was accepted with an amend-ment favoring architectural designs "consistent with the primary conserva-tion purpose of the Seashore."[185] Unhappily for him, the superintendent would be required to make the ultimate architectural judgments.

In reviewing the first draft of the suggested guidelines for the secretary of the interior at the December meeting, the matter of architectural stan-dards surfaced again. Why should there be a double standard for public fa-

cilities such as the visitor centers, Barbara Mayo asked. Park Service buildings are permitted to depart from traditional architecture if they blend in
well with their surroundings, but not so a private dwelling. The way the
committee report was first written, Dexter Keezer rejoined, a dwelling
could look like a weiner or a milk bottle![186]

The other major management issue also came before the Advisory Commission at its December 1977 session. After more than a year of study, four
major public meetings, consultation with the Solicitor's Office of the Department of the Interior on jurisdictional questions, and the active assistance of
scientists from the regional office, Barbara Mayo's subcommittee on ponds
management had rendered its final report. In a document Superintendent
Hadley described as a "remarkably complete, thoughtful, in-depth prod-
uct,"[187] Dr. Mayo asserted that cultural eutrophication threatened to eliminate in a matter of decades some ten thousand years of existence for the
Truro/Wellfleet kettle ponds. Her subcommittee offered six specific recommendations for the management of the freshwater ponds lying within the
Seashore, including the development of tailored management plans for
each pond plus a cession of concurrent jurisdiction over the ponds by the
commonwealth of Massachusetts. Chairman Nickerson remarked that the
ponds management study illustrated once again how much of a resource
the Advisory Commission was for the Cape as a whole.

Tempering the year-end discussions were two other issue-echoes from the
past. Chairman Nickerson, with some personal satisfaction, reported a letter
from William J. Whalen, the new director of the National Park Service, laying to rest the erosive matter of a recreation area category for the Cape Cod
National Seashore.[188] The administratively created management categories
had all been abolished. The second matter was the forthcoming expiration of
the Advisory Commission's two-year charter on December 31, 1977. Superintendent Hadley assured the members that every effort was being made to
expedite the nominations and appointment processes; but as *The Cape Codder*
editorialized, it was entirely possible that the Advisory Commission could
become the "innocent victim" of the Carter administration's "worthy drive"
to reorganize and reduce the size of the federal government.[189]

Overshadowing all of these business matters was a personal event of
some moment to the veteran members of the Commission. They learned
that their ranks would be reduced by one. For reasons of ill health, Ralph
Chase would not be taking his accustomed seat as Eastham's official representative, although he had been asked by the selectmen to remain as a special town consultant on Seashore affairs. A special resolution of the Advisory Commission paid genuine tribute to "Eastham's wise choice" over the
years,[190] applauding Chase's special insights into discussions involving
Seashore and community interests.

It was not until August of 1978 that the Advisory Commission could meet again. Superintendent Hadley apologized for the delay in reactivating the Commission, but much of the difficulty rested with the nomination process. As *The Cape Codder* observed, the Advisory Commission had been the "victim of bureaucratic muddle"[191] — first a lag at the federal level and later delayed nominations from the governor of Massachusetts. There had been some interesting changes in the charter too.[192] In subtle language modification, the Commission would henceforth report to the superintendent, not merely file its reports and minutes with him. In two years' time, the Advisory Commission had gone from a "creature" of Congress, to a "creature" of the secretary of the interior, to a "creature" of the superintendent![193] The experiment with subcommittees had been so successful that specific provision for ad hoc committees "for special purposes," subject, of course, to the general provisions of the Federal Advisory Committee Act, had been added to the charter.[194] And to ensure that the Advisory Commission would enjoy a full term, the new charter carried no specific date for termination — rather a stated period of two years from the date of filing in the Federal Register.

Once officially in business, the Advisory Commission lost little time in getting to work. The most significant issue was the aftermath of the great storm of February 6-7, 1978, which had washed away most of the Seashore facilities at Coast Guard Beach in Eastham, eliminated cottages on Nauset Spit (including the famed Outermost House of Henry Beston), and breached the dikes and dunes at Hatches Harbor and Long Point, Provincetown. In the intervening six months, the National Park Service's Denver Service Center had prepared an assessment of alternatives. Temporary shuttle buses had been placed in service between the Doane Rock and Salt Pond Visitor Center parking facilities and the ocean beaches. More than $2 million would be required to replace the Coast Guard Beach facilities, the Denver planners estimated. Mrs. Laura Underhill, chairman of Eastham's Board of Selectmen, told the Commission that her town was genuinely appreciative of the Seashore's efforts but did not regard busing as the ultimate answer for what she termed "our life's blood," Eastham's summer tourists.[195] Indeed, preliminary statistics revealed a drop in beach visitation of over two-thirds for the summer season. The public preferred the town beach at Nauset where parking was available to the three-hundred-yard walk to the Seashore beaches in Eastham.

The Advisory Commission was also informed that the concession leases on the two Seashore-owned motels, Nauset Knoll Motel in Orleans and the Salt Pond Motel in Eastham, were coming up for renewal. This was a difficult issue for the Seashore, an unwitting owner of these properties due to forced acquisitions early in the life of the park. The matter was equally sensitive to the Advisory Commission, for private motel owners in the area were increasingly vocal about unfair competition from the government. A special subcommittee headed by Chatham's David Ryder, the newly elected chairman of the Advisory Commission,[196] brought in a report recommending a five-year renewal for each motel. At the end of this period, the Salt Pond Motel buildings should be removed, historic Salt Pond House committed to more appropriate uses, and the site returned to its natural state. The Nauset Knoll Motel was a different story. It might become a district headquarters for the Seashore at the end of the permit period, or the commercial permit could even be renewed again. In any event, the Ryder subcommittee maintained, it was time to fix a termination date for each of these properties. There was no disagreement from the commission, although Truro's Dexter Keezer did observe that it was "positively wicked" to pull down needed accommodations.[197]

The November meeting of the Commission revealed a number of changes in the wings. A new Orleans representative, Edward J. Smith, had joined the Advisory Commission, replacing Linnell E. Studley, the selectmen's earlier nominee. Lack of an Interior representative continued to be vexing. Even more significant was the unwelcome news that Lawrence Hadley, the Seashore's superintendent for the past five years, would retire early in 1979.[198] His replacement would be Herbert Olsen, a historian by profession and a veteran of superintendencies at the Boston area historical parks and other locations. Olsen had first come to know Cape Cod through the joint meeting of the Minuteman and Cape Cod advisory commissions held in 1963. His recent experience had been as deputy superintendent of the complex urban park at the Gateway National Recreation Area in New York.[199]

To the continuing discussion over the replacement of Seashore facilities at Coast Guard Beach in Eastham, another issue area was added. What was being done about the migrating dunes at the Provincetown end of the Seashore, Barbara Mayo asked?[200] Research is certainly needed and possibly stabilization through planting and reforestation. She was assured that the topic would be on a future agenda.

At the Advisory Commission meeting of January 19, 1979, a new and different Interior designee was on hand. She was Dr. Sally H. Lunt, psychotherapist, feminist, and the active Democratic vice-chairman of the National Women's Political Caucus. A lifetime summer visitor to Wellfleet, Dr. Lunt was nonetheless watched carefully by the Commission's veterans, ultimately winning their confidence by the distinctive hats she wore, her Cape Cod practice of eating pickles with her chowder, but mostly by her incisive and forthright analysis of issues.[201]

At this same meeting, Commission members learned that the leak from the Amoco gas tank in Truro, which threatened Provincetown's South Hollow wellfield, would require extensive attention, thus requiring a continuation of the special permit for an emergency well within the Seashore. Also, the Advisory Commission received a complaint from Provincetown resident Mark Primack stating that there was no place to enjoy nature within the former Province Lands free from either beach or vehicle use. He suggested setting aside Wood End as a vehicle-free area. Commission member Clifford White,[202] a former vice-president of the Massachusetts Beach Buggy Association, raised an immediate objection to the elimination of a prime fishing area. The Advisory Commission agreed to invite Primack to the next meeting and hear the issue out.

True to its word, the Commission gave thoughtful attention to an essay Primack had prepared entitled "ORV's, Man, and Nature in the Provincelands," an area he described as the "soul of the creative individual."[203] Primack advised the Commission of Provincetown's recent town meeting vote in favor of limiting beach buggies in the Province Lands and the spate of recent editorials favoring the elimination of oversand recreational vehicles entirely. "Thoreau walked here," he observed, "and artists and writers, such as Eugene O'Neill, have traditionally come here to seek inspiration."[204] Upon motion of Joshua Nickerson, the Advisory Commission voted to endorse the principle of vehicle-free areas in the Province Lands, but it also elected to establish a special subcommittee under the chairmanship of Clifford White to explore the application of the principle in practice.

The Commission also learned of a new development in the matter of the Eastham beaches.[205] Continuing vandalism had forced the filing of a declaration of taking by the National Park Service for the Dahlberg property near Coast Guard Beach, an eight-acre tract already so modified from its natural condition that it might become a parking facility for the new beach development. The declaration of taking was a step beyond condemnation, permitting immediate possession of the property. It was time, the Commission concluded, to constitute a special subcommittee to work cooperatively with Park Service planners on the development options for Coast Guard Beach. Elizabeth Worthing, Eastham's representative on the Advisory Commission, was the logical choice to chair the subcommittee.

By the June 1979 meeting of the Advisory Commission, the Worthing subcommittee had begun to untangle some of the options. Mrs. Worthing reported that Eastham was not entirely adverse to an additional land exchange with the Seashore to expedite the development program, but that some of the lands the town desired, such as additional land for municipal purposes near Salt Pond, could be troublesome to the Seashore.[206] But the primary difficulty was the cumbersome planning process followed by the National Park Service and the preliminary nature of its plans and commitments. Eastham was fast becoming resigned to near-permanent shuttle bus service to its beaches.

Later on in the meeting it became evident from the White subcommittee report that the Province Lands vehicle-use issue was a bear by the tail. An open meeting held at the Salt Pond Visitor Center on May 17, 1979, had attracted 150 participants. One thousand signatures had been submitted opposing restrictions on oversand vehicle usage. Nevertheless, the Advisory Commission decided to see for itself. An all-day field trip to the Province Lands was scheduled for August 24. Commission members encountered evidence of ORV violations and heavy foot traffic from the dunes parking lot on Route 6 clear to the ocean. The barrier beach at Long Point was still virtually an island due to the February 1978 storm, but the break at Wood End had brought about a healthy rise in tern productivity. The Commission found the breach at Hatches Harbor on the mend thanks to sand accretion and natural revegetation. It was evident that reduced vehicle usage had contributed to the recovery.

For the October meeting of the Commission, Superintendent Olsen arranged a presentation on the five years of research conducted by the Park Service's Cooperative Research Unit at the University of Massachusetts, Amherst.[207] Dr. Paul Godfrey, a respected botanist, had been carrying out the work under the general supervision of Dr. Paul Buckley, regional scientist for the North Atlantic region. Dr. Godfrey described the issue of ORV use as very emotional. Nevertheless, his research had led him to conclude that much of the dune migration was attributable to indiscriminate driving. He cited the so-called drift line at the margin of the ocean shoreline as particularly sensitive, because it was the source of natural materials that would start the process of revegetation. He also indicated that a complete ban should be placed on the salt marsh areas and tidal flats because of the risk of permanent damage. The intertidal areas, where constant physical change is taking place anyway, and carefully selected routes through the inland dunes, were likely to display the greatest tolerance to vehicle usage. Dr. Godfrey stopped short of recommending a complete ban on oversand vehicles, but his report clearly favored substantial modifications in past practices. Dr. Sally H. Lunt, the Interior's new designee on the Advisory Commission, spoke for Chairman White in presenting the special subcommittee's report on ORV usage. The subcommittee recommended creation of a limited vehicle-free area on the outer beach, restricted during daylight hours only, permanent closure of the isolated area at Hatches Harbor, and stepped-up enforcement by Seashore rangers to prevent further damage by "dune busters." It was agreed that the ORV issue would be the centerpiece for discussion at the December meeting.

The opening salvo was fired by Dr. Barbara Mayo, cofounder of the Center for Coastal Studies in Provincetown and, paradoxically, Clifford White's counterpart as a commonwealth of Massachusetts representative on the Advisory Commission.[208] The vehicle-free area specified is not big enough, she said, and the restriction should be in force for a full twenty-four hour period. Moreover, any action should be preceded by a thorough study of oversand vehicle usage throughout the Seashore as a means of "ridding biases." Truro's Dexter Keezer spoke supportively. "The talk is over people and vehicles . . . not what is happening to the environment," he said.[209] The Commission was reminded that concern over the effects of vehicle usage was not new; the 1893 report of the trustees of Public Reservations on the Province Lands, for example, had blamed cart ruts for much of the dune destabilization. Clifford White placed most of the onus on a few thoughtless ORV operators. This triggered a long discussion about the adequacy of Seashore enforcement. It was agreed that stepped-up educational activity, increased patrols, and closure of areas that were unenforceable should be pursued actively by the National Park Service. Numbers are the real problem, Joshua Nickerson observed.

The Advisory Commission then proceeded to vote on each of the eleven subcommittee recommendations. In a split vote (two members opposed), the Commission approved Dr. Mayo's major modifications favoring the establishment of vehicle-free areas all the way from High Head to Head of the Meadow Beach, and from Herring Cove Beach south to Long Point, on a twenty-four-hour-a-day basis. However, it rejected her suggestion that the inland trail through the dunes be closed completely to ORV use. A majority of the Commission also agreed that the Seashore should be asked to provide an assessment of its enforcement capabilities in the Province Lands and elsewhere. Despite her strong feeling that ORVs should eventually be phased out, Barbara Mayo admitted that the Commission field trip had made her feel "less antagonistic" to this special-use group and "more positive about that kind of activity than before."[210]

Before the meeting closed, Superintendent Olsen reminded the Commission that a revised land acquisition policy had been adopted by the National Park Service and that the Cape Cod National Seashore (among other park areas) was required to prepare a formal land acquisition plan. He reported that most of upland included within the original boundaries had already been acquired by the Seashore. Less-than-fee approaches, such as the scenic easements obtained on Hog and Sampson islands in Pleasant Bay, would be encouraged in the remaining cases. He invited the Advisory Commission to work with the Seashore in preparing the required plan. It was agreed that a special subcommittee would be constituted for such purposes under the chairmanship of Joshua Nickerson.

By the March 1980 meeting of the Advisory Commission, the matter of oversand vehicle usage had clearly become a dominant issue.[211] Sportsmens' groups had demanded time to present their case. Waiting in the wings was the Tranquility Lobby,[212] a self-styled citizen group in Provincetown equally anxious to present its views. Kenneth Dutra, chairman of the Citizens' Right to Access Committee, was the leadoff spokesman at the March session, claiming to represent over sixty-five thousand concerned sportsmen. He protested the Commission's vehicle-free recommendations and closure proposals, stating that the Commission had no right to change the fishermen's way of life because they had done no damage to the ecosystem. He was supported by Elizabeth Lackey who observed that ORV damage is negligible compared to the damage caused by Mother Nature. Matthew Costa, president of the Highland Fish and Game Association in Provincetown and a dune tour buggy operator, was more outspoken, claiming an outright conspiracy against ORV users. The sportsmen cited the National Park Service's own study report which did not recommend banning oversand vehicles. They held the Advisory Commission, not the Seashore, really responsible for the closure recommendations and served notice that the Massachusetts Council of Sportsmens Clubs would seek an appointment to the Commission in the near future.

Dr. Herbert Whitlock, spokesman for the Association for the Preservation of Cape Cod, called the discussions one-sided. He claimed that the scientists' findings had been taken out of context. Dr. Mayo reminded the generally hostile audience that the lands need to be managed in a way that is best for everyone. "There is only one Cape Cod National Seashore," she said.[213] Commission members disagreed on whether the ORV issue had been heard fully. The consensus was that other interests should be allowed to speak, and, at the least, Commission members should review carefully the individual statements submitted and notify the superintendent of their feelings in the matter.

The remainder of the session was devoted to a discussion of the draft land acquisition plan for the Seashore.[214] Nearly fifteen thousand acres were likely to remain in federal, state, or local ownership—much of them submerged lands under the jurisdiction of the commonwealth of Massachusetts. Civil action was pending on 1,161 acres, the highest priority for government acquisition. A three-man panel of retired judges was, at long last, expediting these cases. Less than a hundred acres of unimproved land, buildings constructed after the 1959 cutoff date, and properties in violation of zoning provisions remained to be acquired. As of February 8, 1980, the report showed 582 individual tracts within the Seashore in the category of "improved properties." The matter of the shacks on the dunes in Provincetown and Truro, a headache conveyed to the National Park Service along with the earlier state transfer of the Province Lands and Pilgrim Spring State Park, remained a matter for administrative resolution. After all, Nicholas Wells of Provincetown observed, "one man's shack is another man's castle."[215]

By the May meeting, Superintendent Olsen could report that the regional director was reviewing carefully the Commission's ORV closure recommendations.[216] It was likely that the action would be construed to fall under the provisions of the National Environmental Policy Act and require a full environmental impact statement. If so, it would be prepared by the regional compliance staff with inputs from the Seashore. Dr. Mayo expressed her distress at the superintendent's apparent inability to close an area when obvious environmental damage was being done. Dr. Lunt felt equally dismayed that only now had it been found necessary to conduct another study. Joshua Nickerson spoke of the bureaucratic necessities that seem to avoid decision making. He hoped the situation would not get worse. In point of fact, it had already deteriorated because the Provincetown selectmen, based upon a substantial town meeting vote, were strongly opposing the recommended ORV closures at Long Point and Hatches Harbor.

The Advisory Commission had preceded its business meeting with a fact-finding visit to Provincetown Airport, where a new instrument landing system had been proposed. With modifications, the development was found to be compatible with Seashore objectives and was recommended favorably to the superintendent.[217]

By September of 1980, the Advisory Commission was at full strength and duly sworn in for its next two-year term. David Ryder of Chatham was reelected chairman. Dr. Sally H. Lunt, the secretary of the interior's representative, replaced Provincetown's Nathan Malchman as vice-chairman, who received a special vote of appreciation for his long service as member and officer. Sherrill B. Smith had become the new representative from Orleans. Superintendent Olsen used the meeting to update the Commission on pending matters.[218]

He reported that the proposed ORV management revisions had indeed gone the route of an environmental impact analysis. As part of that process, public meetings would be held in Faneuil Hall, Boston, and on Cape Cod in November. The National Park Service hoped to make a decision by February and place the selected management option in effect for the 1981 season.

With respect to the Eastham beach development program, the alternative recommended by the Advisory Commission had been selected by the regional director on August 28, and funds for construction would be requested in the budget starting October 1, 1982. Joshua Nickerson remarked that it had now been five years since an act of God destroyed Coast Guard Beach. He found it astonishing that even temporary facilities could not have been provided in the interim. Eastham's Elizabeth Worthing observed that, under the circumstances, it was becoming increasingly difficult to deflect the residents' desire for a town-owned beach. Orleans's Sherrill Smith reported that the lack of beach facilities in Eastham was putting unbearable pressure on his town-operated Nauset Beach. Regional director Richard Stanton, present for the meeting at the request of Superintendent Olsen, could offer no encouragement. Virtually the entire 1981 construction budget for the North Atlantic region had been wiped out, he reported, and the National Park Service had no more contingency monies available.

Despite these management frustrations, the Seashore's programs were basically healthy. Visitation was up 14 percent for the comparable period, and the interpretive programs, including the new demonstration cranberry bog and trail, had proved unusually popular. A Youth Conservation Corps camp with twenty-three enrollees had cooperated with the Seashore and the town during the summer under the theme "Work hard, play fair, nobody hurt!"[219] An Urban Initiative Program was bringing children from Boston's inner city to Cape Cod three days a week. Moreover, at the suggestion of the Advisory Commission, the Seashore had renewed its arrangement enabling the Nauset Regional School District to continue to use the former Job Corps Camp facilities at Camp Wellfleet for educational purposes.

By the July meeting of the Advisory Commission, Superintendent Olsen could report that the Provincetown Airport safety improvements were proceeding satisfactorily.[220] A waiver of FAA requirements would be sought to preserve the remaining stand of pitch pine in the Race Run section of the Province Lands. The instrument landing project was estimated to cost some $500,000, 75 percent of which would be provided by the federal government. Olsen advised that Provincetown's was the only airport on the Cape and islands without these navigation improvements. Airport manager John Van Arsdale assured the Commission that he and the National Park Service were in agreement on the steps that needed to be taken.

The first meeting in 1981 was marked by a moment of silence in memory of two Commission members—Ralph Chase of Eastham and John Whorf of Wellfleet—who had passed away during the intervening months.[221] This was also to be Marjorie Burling's last meeting before retirement. As personal secretary to the superintendent, Mrs. Burling had taken the minutes of ten-and-a-half years of meetings and had "worn out" four superintendents. There would also be a new designee of the secretary of the interior, it was reported. Paul F. Nace, Jr., a former associate commissioner of the Metropolitan District Commission in Boston, had been selected to replace Dr. Sally H. Lunt. And it was now time to review the hearings record of the ORV public meetings for possible modifications to the management plan recommended earlier to the Seashore.[222]

Joshua Nickerson opened the debate by offering a prepared statement. He favored phasing out entirely the overnight uses permitted self-contained ORVs, arguing that continued use simply invites similar use by nonfishermen. Elizabeth Worthing commented that the fishermen presently have a special privilege not enjoyed by other Seashore users. Clifford White and Sherrill Smith preferred to settle for limiting self-contained ORVs to their present numbers. Nickerson felt this would be grossly unfair. Barbara Mayo offered a motion to ban all such vehicles, which was voted down by the Advisory Commission. Provincetown's Malchman observed that it would be unrealistic to try to protect every square mile of dune, as it were, by enclosing them in Saran Wrap. A substitute motion to minimize ORV use, not expand it, won the endorsement of a majority of the Commission. By an even narrower margin (four yes, three no, two abstentions), Nickerson's motion to eliminate residential occupancy of oversand vehicles on a planned, long-term basis was approved by the Commission.

Barbara Mayo returned to her earlier advocacy of extensive vehicle-free areas and reduced inland trails. Her motion was amended to retain the suggested vehicle-free area at the northern end of the ocean beach, but the stretch from High Head to Head of the Meadow would be vehicle-free only during daylight hours and summer months. On the question of vehicle numbers, the daily limit was believed to be more significant than a seasonal limit. Seashore rangers were emphatic that the enforcement of numbers and scattered closure areas could become an administrative nightmare, but they were willing to give any reasonable solution a try. Truro's Dexter Keezer stepped in with some timely compromise language: the number of permits issued should be set by the limits of enforceability and be determined administratively by the Seashore.

But the aspect commanding the most attention was the public meeting testimony delivered earlier by David B. H. Martin. Martin, former legislative counsel to Massachusetts senator Leverett Saltonstall and the chief draftsman of the Seashore legislation, claimed that the National Park Service was legally in error in granting any ORV permits because vehicle use had been scientifically confirmed as detrimental to the Seashore. His friend and former colleague, Jonathan Moore, now the director of the Institute of Politics at Harvard University's John F. Kennedy School of Government, also testified. Moore suggested a phasing out of ORV usage over a five-year period as a more pragmatic course to take.

By the April meeting of the Advisory Commission, another facet of Province Lands management was ready to be addressed—a belated response to Barbara Mayo's earlier inquiry about the prospects of dune stabilization.[223] Dr. Stephen Leatherman of the Cooperative Research Unit in Amherst had been asked to look into the matter. In an oral progress report to the Commission, he traced the history of erosion control efforts back to an act of the colonial legislature in 1715 forbidding the grazing of livestock and the cutting of wood. Although the large, migrating dunes are now part of the accustomed landscape, Leatherman reported, the Province Lands were once heavily forested. The remaining beech forest, just a remnant of the original hardwood forest, is being buried by sand at the rate of one-half acre per year, he said. Mt. Ararat, the highest elevation in the Seashore, is losing fifteen feet of sand each year due to uncontrolled erosion, much of it filling in nearby Pilgrim Lake, the spectacular reflecting water body for the high dunes. Dr. Leatherman promised the Advisory Commission a complete report on the likely management options, but his preliminary conclusion was that a process of revegetation, much like the program undertaken by Province Lands superintendent Small during the 1800s, could prove successful.

The Wellfleet portion of the Seashore was the next topic for discussion. A proposed landfill, wholly on town property but surrounded on three sides by Seashore ownership, was stirring local concerns. Town Counsel Charles E. Frazier, Jr., an early antagonist of the Seashore, wondered why the Seashore and the Advisory Commission were involved. He reminded the Commission that it was he and Judge Francis Biddle who had insisted that an advisory commission be appointed to protect the interests of the communities, but he hardly expected the Commission to intervene where it had no official business. Frazier had been assured that the matter was an information item only prompted by questions raised locally, but there was some evidence that the regional office of the National Park Service was prepared to construe its general statutory authority broadly where an activity even outside the Seashore might impact adversely on its interests. The matter of Wellfleet's sanitary landfill, located as it was on the bay side of the Seashore, directed the Commission's attention to the status of the extensive Herring River marshes, which were 70 percent in Seashore ownership. This topic would not take long to surface again.

By the June meeting of the Commission, the Truro portion of the Seashore was up for detailed discussion.[224] The Advisory Commission's patience was wearing thin over Provincetown's continuing requests for emergency use of Truro's South Hollow wellfield. Provincetown town manager William A. McNulty and Camp, Dresser & McKee consulting engineer Robert Weimar were asked to present the particulars. Engineer Weimar took the Commission step by step through a thorough discussion of the geology and hydrology of the lower Cape and described Provincetown's continuing efforts to prevent the entry of suspected carcinogens into the town's main water supply. A two-to-three-thousand gallon gasoline spill adjacent to Route 6 had not yet reached the wellfield, but only because the cone of depression caused by normal pumping had ceased when the well was shut down. Since that time, Town Manager McNulty reported, Provincetown had adopted stringent water restrictions, metered the entire community, and explored alternative sources. The best solution appeared to be a cleanup of the spilled material, now extensively underground, so that the well could be put back into production. In the meantime, the town had to depend upon emergency supplies from a location within the Seashore made available under special permit.

Robert Weimar then described the options available to the town and the likely costs. A conventional, activated-carbon treatment plant would cost upward of $1 million and would require a point of discharge into the ocean. A second option was the use of bioreclamation techniques through the injection of bacteria to literally "eat" the hydrocarbons, a proven but less conventional approach that might cut the costs in half. However, the regulatory agencies and the local public had expressed some uneasiness with this technique. Weimar estimated that either facility might take eighteen months to construct and would need to operate for a minimum of five years. If no cleanup action was taken, natural drainage would discharge the first contaminant into Cape Cod Bay in twenty to forty years' time. Hundreds of years might be required to fully cleanse the aquifer. The effects on the bay itself at that time were matters of speculation, but a continuing sheen of oil of considerable duration could be expected.

In a lively question and answer session, Provincetown water superintendent Paul Daley stated emphatically that Truro would not permit bioreclamation because of a fear of bacterial contamination. In contrast, Dr. Herbert Whitlock spoke of the experience of the Association for the Preservation of Cape Cod in the Woods Hole area, stating that bioreclamation seemed both safe and best for the problem. He put Provincetown on notice that his association would oppose in court any ocean discharge. Commission member Barbara Mayo asked whether the treatment plant effluent could be injected at the ocean shoreline. The consulting engineer replied in the affirmative but noted that Seashore permission would be required. Superintendent Olsen asked whether Provincetown's entire supply could be treated, thereby removing the necessity of a temporary treatment facility at South Hollow. The answer was again yes, but the option would be extremely costly. Weimar added that the special treatment facility, once its primary objective had been satisfied, could then serve as a backup source of supply, something that Provincetown lacked and needed. Town Manager McNulty informed the Advisory Commission that he had requested a three-year extension of the special permit to enable Provincetown to install the required facilities. Without dissent, the Commission voted its encouragement of a cooperative water supply approach by the Seashore and the affected towns.

On August 7, 1981, the twentieth anniversary of the creation of the Cape Cod National Seashore, a special ceremony was held at the Salt Pond Visitor Center.[225] The setting was a cloudless Cape Cod day, in the outdoor amphitheatre at Salt Pond. Several hundred local residents, summer visitors, and dignitaries were on hand. Superintendent Herbert Olsen opened the morning ceremonies with welcoming words and introduced the principal speaker, David B. H. Martin, who made appropriate remarks on the origins of the Seashore.

Martin spoke of his "enormous pleasure [and] considerable pride" at performing these honors and recalled the occasion twenty years ago, at exactly twelve noon, when President John F. Kennedy, with twenty-two different pens, signed the act creating the Cape Cod National Seashore. Martin presented to Superintendent Olsen the pen he had been given at the ceremony, and a facsimile of the title and signature pages of the document bearing the "sprawling signature" of Speaker Sam Rayburn and the "tight, firm signature" of President of the Senate and Vice-President Lyndon B. Johnson. But Martin's reflective remarks about the accomplishments of the Seashore, and its Advisory Commission, appropriately closed out this second decade of activity.

In 1961, Martin asserted, Cape Cod was facing a potential "tragedy of the commons." Its open heritage of ocean beaches and adjacent uplands was being ungulfed by visitors and development pressures. Passage of the Seashore Act seemed to have illustrated the "genuine benefits" to be gained through government and laws, he said. In Cape Cod's case, the building blocks of the law were "unique and distinctive," Martin observed, for the legislation contained a number of special features. For example, it specified the boundaries of the Seashore in the interest of the people living there. It provided authorization for the appropriation of funds for acquisition. It suspended the secretary of the interior's power of condemnation after town zoning action, thereby ensuring stable town tax bases, saving public acquisition funds, and asserting town control over land use. It included provisions barring any development affecting unique flora and fauna. And it created an Advisory Commission "carefully crafted to be a mechanism through which the tastes, the inclinations, the manners, the preferences, the disinclinations of everyone affected by this legislation would have a vehicle for expression and effect."

In preparing his remarks, Martin had set about to read nearly four thousand pages of Commission minutes. He found them to be a "repository of historical detail" and was impressed by the extraordinary diversity of the agenda subject matter. In his view, the Commission was "never out of touch with social reality"; it had "more than fulfilled the highest hopes and expectations" of the act. The words of the legislation had been transformed into a "living experience"; it had enlisted "legions of participants." Your conduct, he told the Advisory Commission and its assembled audience, has and should serve as a model for imitation in other public-use areas throughout the country.

The Cape Cod Experience

On August 7, 1981, the Cape Cod National Seashore was twenty years old. Nearly seventy million visitations had been made to its developed beaches, bicycle paths, interpretive trails, and visitor facilities. By this date, the federal government had spent more than $40 million on land, $15 million on facilities, and $20 million on operations—an investment of approximately one dollar per visitor day. Another $75 million had almost certainly been generated in local expenditures by Seashore visitors. Physically, the Seashore was almost complete. More than twenty-eight thousand acres of fresh water, upland, ocean beach, and tideland now lay within a day's drive of a third of the nation's population. Yet incredibly, one could still climb the dunes at High Head in Truro or Marconi Beach in Wellfleet and, as Henry Thoreau declared, "put all America behind."[226]

There is a lesson here for all Americans, struggling to preserve a remnant of their natural heritage in the face of relentless pressures for growth and development, for among several "firsts," Cape Cod was the first national park to become established in an area that had been settled and civilized for over three hundred years. How that goal was accomplished and what it means for the future are the objectives of this final section.

THE AUTHORIZATION PERIOD

The authorizing legislation for Cape Cod was not won without a struggle. A combination of factors contributed to its ultimate passage. On the congressional side, sponsorship by two influential senators, one of whom (John F. Kennedy) was shortly to become president of the United States, gave strength and stature to the undertaking. Senators Saltonstall and Kennedy had enviable records of bipartisan action on behalf of their commonwealth,[227] and they were supported strongly by other members of the Massachusetts delegation, notably Congressmen O'Neill and Boland who themselves commanded key sectors of influence within the Congress.[228] Leadership by the two senators had another important effect. It liberated Congressman Hastings Keith to devote his best efforts to representing the interests of his Cape Cod constituents.[229]

Even more significant than the seniority and stature of the two Massachusetts senators was the total commitment of their respective staffs to the Cape Cod legislation. The cause became a crusade, and the crusade led to one of the most painstaking and innovative legislative vehicles in national park history.

Yet, Cape Cod was helped measurably by events nationally. The National Park Service, a professional agency whose credentials were respected by both the public and the Congress, was on the threshold of a major effort to preserve the nation's shoreline, of which the Cape Cod seashore was only one example.[230] Buoyed by the national groundswell of participation in outdoor recreation, there were significant blocs in Congress prepared to back seashore legislation. Both on its own merits and as a precedent for other areas, Cape Cod could be assured of substantial support, particularly from eastern representatives who could justifiably raise questions of inequity in the preponderance of national parks in the West.

Cape Cod had another plus going for it politically—its visibility. Every schoolchild knew about Cape Cod and the Pilgrim settlement of New England. It was instantly identifiable as a national resource in contrast with Oregon Dunes in the West or Padre Island on the Gulf Coast. As a vacation resort, Cape Cod had already been visited by countless thousands of people, and many had seen firsthand the creeping commercialism of which the Seashore proponents made so much. Besides sheer numbers of supporters, Cape Cod had become a mecca for many of the cultural and intellectual elite, who commanded a disproportionate degree of influence in political and journalistic circles and were not afraid to use it.

On Cape Cod itself, there was also a substantial critical mass ready to support reasonable seashore legislation. All but the unenlightened were well aware of the mounting development pressures, and the native Cape Codder was not yet prepared to sacrifice a heritage of unspoiled ponds, beaches, and woodland for short-term economic returns. Here the New England town meeting system displayed its real value, for there was a recognized and tested framework for the conflict and controversy that would mark the Seashore proposal over nearly two years' time. By the end of it, public understanding would grow and a way would be found for acceptable compromise.

Today, most previous opponents of the Seashore still will not admit to a change of heart, but the fact remains that the park is an accepted part of their communities. There is a two-fold reason for this viewpoint. First, most local observers believe the Seashore has done an excellent job of operations and development. And second, the park has turned out to be better than most people were led to believe.

THE OPERATIONAL PERIOD

In terms of its administration, the Cape Cod National Seashore exhibits many characteristics worthy of prototype status. It has opened the eyes of millions to the fragile and sensitive landscape of Cape Cod through a particularly effective interpretive program. It has utilized the landscape diversity of Cape Cod to enable visitors to enjoy both active and passive uses of the Seashore. Its developed facilities are modest and in good taste, distinct but not intrusive. There is a natural feel to Cape Cod despite its capacity to absorb the second largest visitor load in the North Atlantic region. The net result of many hours of discussion and debate, and many versions of master plans, has been a careful blend of conservation and recreation that more than fulfills the intentions of the legislative sponsors and the determination of its Advisory Commission.

In terms of operation, Cape Cod has set a high standard for other jurisdictions. The stable land base represented by the Seashore has encouraged private investment and reinvestment both outside its boundaries and for those properties remaining within but subject to protective zoning. It has acted as a magnet for retirees, thereby enriching the cultural environment of the lower Cape and providing the Seashore communities with an economic base that maximizes tax returns and minimizes service expenditures. There is little question that the Seashore's presence has helped Cape Cod toward its long-range goal of an extended visitor season and a viable year-round industry.

Cape Cod also illustrates that a national park can enjoy an effective and even symbiotic relationship with adjacent communities. The tradition of candor and openness is noteworthy. It has paid important dividends in terms of public and community relations. A working advisory commission, which enjoys the support of both the National Park Service and the participating jurisdictions, serves as a window to the outside world, thereby preventing the sense of isolation and parochialism that characterizes so many large national parks. Cape Cod can serve and has served as a resource to fulfill a wide variety of public needs—from the national Job Corps and Environmental Education programs to the supplemental police and fire protection services so essential to the surrounding towns.

One could argue that good relations with the towns became more of a necessity than a sought-after goal. The towns of Provincetown, Truro, and Wellfleet, for example, were so squeezed land-wise by the Seashore that their cooperation was virtually mandated. Yet there was a deliberate effort by the National Park Service to achieve a close working relationship with the community, which appears to have succeeded.

In that respect, the National Park Service itself has learned a few lessons. Prior to Cape Cod, the Service was conditioned to the role of custodian for many square miles of wilderness and natural terrain. It was the undisputed master of all it surveyed. The agency was a disciplined professional agency, accustomed to making decisions in a hierarchical, bureaucratic manner. It was utterly unprepared for the participatory and at times skeptical democracy of Cape Cod. Yet the career nature of the Park Service, and the obviously precedental character of the new Seashore, enabled it to adjust. Even the top administrators of the Service had risen through the ranks. Most had been park superintendents themselves and fully appreciated the independence of judgment and the necessity for delegated authority in such positions.

There is evidence that the National Park Service determined to do everything within its power to make Cape Cod a success, even to the point of bending or making precedent. For as the northeast regional director Daniel Tobin was reported to have told his staff meeting in Philadelphia on the eve of passage of the enabling legislation: "Next to the Organic Act creating the National Park System, the Cape Cod bill is the second most important legislation in park history. There will be a different kind of park management than before. It will be difficult but we will do our best to live up to it."[231]

The Seashore did stumble occasionally. Its master planning, predicated as it was on natural and scientific features, did not always fit the tenor of Cape Cod. Though sensitive to conservation considerations, the lower Cape was equally concerned with social and economic conditions, but here again the extraordinary Cape Cod instrument and philosophy prevailed. Under the leadership of the Advisory Commission, the early master plan drafts were scrapped and redrawn. Unconsciously, Cape Cod was pioneering the open planning approach so much in vogue today. Largely because of that process, the Seashore remains one of the earliest areas in the entire national park system to develop a complete and viable master plan.

In recent years, the Seashore's development program has tended to limp along. The sequential development planning process, administered from the distant Service Center in Denver, has not been without its problems. Once Cape Cod lost its special luster in the National Park Service firmament, it fared no better than any other park in the authorization and appropriations process. A good case was the aftermath of the great storm of February 1978, which washed away virtually all of the developed beaches in Eastham. Five years later, the Seashore was still making do with temporary facilities. And democracy had become a tortuous path to follow. Public meetings were now required at virtually every step along the way. The environmental impact statement had become a weapon for the opponent or even the skeptic. The prospect of litigation could never be dismissed entirely.

To further complicate operational matters, the Seashore was down to the most difficult of cases. A good example was the matter of oversand vehicle usage in the Province Lands section of the Seashore. No solution would ever satisfy all the parties at interest. The Advisory Commission's institutional anguish in trying to settle on a compromise recommendation for the National Park Service merely illustrated the constant tension between conservation and public use in any major public park. The situation would get worse rather than better in the years ahead.

LAND ACQUISITION AND ZONING

Operational elements deserving special scrutiny are the provisions governing land acquisition and zoning. One will recall that the traditional acquisition authority of the National Park Service was tempered by a section of the act suspending the secretary of the interior's condemnation powers if the six towns adopted satisfactory zoning. They had all done so. Before turning to the administration of the zoning provisions, the land acquisition program itself merits some consideration. It is not widely known, for example, that the Seashore legislation constituted the first major national park for which acquisition funds were authorized by Congress and, correspondingly, the first major effort of the National Park Service in the land acquisition field. Heretofore, national parks had either been acquired by gift or carved out of the public domain by special act of Congress. Having devised the basic land-use control instrument and placed the process in motion, the Seashore set about the long and difficult task of acquiring the remaining authorized acreage. There was no precedent for a program of this size, so Cape Cod found its own ways of making the program work.

As first superintendent Robert Gibbs described the situation,[232] the Seashore was faced with many different types of land ownership. Federal ownership was represented by the Coast Guard with its lighthouses and lifesaving stations at Race Point, Wood End, and Long Point in Provincetown; Highland Light in Truro; and Nauset Light in Eastham. The Air Force owned and operated an early warning station in Truro. Camp Wellfleet was another military installation of strategic importance to the Seashore. Monomoy Island at the southernmost end was a national wildlife refuge administered by the Fish and Wildlife Service, but also a natural extension of the Great Beach. The state-owned lands were equally critical, especially the five thousand acre Province Lands and the one thousand acre Pilgrim Spring State Park in Provincetown and Truro respectively. State ownership of the tidelands offshore and the so-called Great Ponds (natural bodies of water ten acres or more in size) presented special kinds of land acquisition problems. In addition, all of the towns owned property within the Seashore's authorized boundaries. Although these lands could not be acquired without the specific consent of the towns under section 2 of the act, the future ownership and use of major town beaches, such as the Coast Guard beaches in Eastham and Nauset Beach in Orleans and Chatham, were crucial questions for the Seashore. Quasi-public ownership included the large tracts of land owned by Brown University and the Massachusetts Institute of Technology in Truro, and the Massachusetts Audubon Society's control of Henry Beston's Outermost House and a surrounding thirty-three acres on Nauset Beach in Eastham.

But the immediate problem was the property in private hands. At the time of the Seashore's authorization, there were 660 private homes, 19 commercial operations, and 3,600 separate tracts of land, many simply parcels of subdivisions that had been divided into lots as small as one-twelfth of an acre. The privately owned commercial properties included motels, campgrounds, service stations, cottage colonies, an art studio, restaurants, and professional offices. Fortunately, the authorizing legislation laid out the actual boundaries of the Seashore, but, upon closer examination, there were many anomalies. Some followed contour lines or high-water tidal lines, which were not readily visible on the land. When natural features were not available, compass directions were used. At the time the exterior survey was made, some boundary lines were found to run through buildings, yards, and gardens, to split ownerships in two, and to separate accessory buildings from residences.

The lack of precedent within the National Park Service caused other problems. The initial lands acquisition office operated under the direct control of the regional office in Philadelphia. There was little autonomy or delegated authority either from the region or from Washington until the new superintendent was firmly in place. Despite willing buyers and emergency situations at hand, a laborious process of appraisals, tract mapping, title examination, negotiation, and approval was required. Priorities had to be worked out too—in the Seashore's case, hardships faced by individuals, large contiguous blocks with clear title, and lands where Seashore developments were likely to take place. Preserving the integrity of the area through removal of threats of unsightly developments was also an early priority.[233]

Then, the various means of acquisition had to be sorted out. Owners entering into agreements to sell their property were given four options: outright sale, sale with the owner retaining life-tenure rights to the homesite, sale with the owner retaining homesite rights for a fixed period of years (up to twenty five years), or sale of the unimproved lands with the individual retaining ownership of the homesite in perpetuity under the Suspension of Condemnation section of the Seashore Act (section 4). In addition to outright purchase, lands could be acquired by donation, transfer, exchange, or condemnation. Fee ownership of the land was the usual practice, but the Seashore could and did negotiate for rights in land, such as easements and conservation restrictions.

Problems of poor title, small ownerships, and boundaries plagued the Seashore's acquisition program from the outset. The original "proprietors' divisions" of lands were measured and marked with a blaze on a tree which tended to heal over even if the tree remained. The later conveyances were in generalities, e.g., "bounded by Noah Nickerson on the north and Hezikiah Horton on the south" or "my woodland near Long Pond." The property corners were witnessed in ways now relegated to archaeological history – a rail fence, a ploughed ridge, one or more fist-sized stones, or even a piece of whalebone. A contributing factor to the uncertainty was the loss of records during the fire in the Barnstable County Courthouse in 1828.[234]

Land had been in some families for literally generations. The Seashore was especially sensitive to landowner apprehensions and consciously determined to move only with deliberate speed. It utilized a full spectrum of acquisition techniques from outright condemnation to life tenure and occupancy. It spent literally hours gaining the confidence of landowners before negotiating for purchase options. The end result was a land acquisition program that had exhausted its authorized $16 million by the end of 1978, the consequence of careful procedures and escalating land values. An additional authorization of $17.5 million was provided in 1971. When the second-phase acquisition program was completed in 1974, the Seashore had cost more than twice the sum estimated at the time of authorization. Yet, the investment had earned important dividends. Upon his retirement from the National Park Service in 1968, George Thompson, the man who had taken more land away from Cape Codders than any other person in history, was told that he had won their respect, confidence, and trust.[235]

There is no adequate explanation for the special zoning provisions of the Seashore legislation. The primary draftsman was a young congressional staff attorney with conservation enthusiasm but little real experience. Far from a grand design, the legislation seemingly just happened. As David B. H. Martin has described it,[236] he began with a blank piece of paper and attempted to be as responsive to the concerns being expressed by Cape Codders as to the aspirations of a great national park. The ultimate instrument was forged in the crucible of peer review, public critique, and political compromise. Small wonder that it has survived the test of time.

The provisions contained in section 4 of the Seashore Act were deceptively simple. In the event of a town zoning bylaw, "duly adopted and valid," approved by the secretary of the interior, the government's authority to acquire improved property by condemnation within the Seashore's boundaries would be suspended, as is the case today in the six Seashore communities. The act described an improved property as a "detached, one-family dwelling the construction of which was begun before September 1, 1959," the approximate date the legislation was submitted to Congress. By regulation and practice, the protected property has come to include the dwelling itself, all accessory buildings (such as boathouses), and at least three acres of unimproved land. Issuance of an official Certificate of Suspension of Condemnation enables the property owner to retain or convey that property indefinitely as long as the use remains comparable. More than 500 private properties were in that category as of February 1980.[237]

Here again there was no precedent to work from – only individuals determined to make the concept succeed. One was Ben Thompson, the affable and courtly chief of recreation resource planning for the National Park Service in Washington. He said to his legal assistant, Elmer Buschman, "Let's go," and Buschman went.[238] Buschman, who had never done anything like this before, set to work extracting from the tentative guidelines, the legislative history, and the Seashore Act every provision that had anything to say about the use or development of property within the Seashore. He then measured these provisions against the existing town bylaws and cases where there was state precedent upheld or unchallenged in the courts (the origin of the three-acre lot size, for example). In countless meetings arranged by the tactful first superintendent, Robert Gibbs, the details of zoning bylaw changes were hammered out and ratified by later town meeting action. When it was over, the Park Service officers were relieved and delighted, for they had established a process and a formula that could be applied to other parks.

The concept of enduring private enclaves, protected by approved local bylaws, constituted a master stroke of operational and political conception. It ensured a continuing tax base for the communities, reduced the acquisition expenses of the federal government, and added a cultural conservation objective to the preservation program. More than that, it cut the political legs out from under the opposition by being fully responsive to assertions of community initiative and responsibility. Once the Seashore legislation had passed, the passage of zoning bylaws represented a tangible way for communities to begin to assert their influence over the new park, and a valuable opportunity for the new Seashore administration to demonstrate its genuine commitment to cooperation with the lower Cape communities.

But as the years wore on, the Seashore was forced to come to grips with an uncomfortable reality. Many of the improved properties protected by the special zoning provisions had become havens for the rich. Pragmatic Cape Codders had profited from loopholes in the law. Protected by the Certificate of Suspension of Condemnation issued by the Seashore, their properties enjoyed a ready market for off-Cape owners eager to own a private sanctuary surrounded by National Park Service lands. In some cases, the loopholes bordered on the ludicrous. Whole new structures could be built around an original feature, such as a fireplace, or were permissible if they incorporated original materials such as sheathing or boards. The Advisory Commission's thoughtful examination of the problem in 1977, and its subsequent recommendations to the National Park Service,[239] are deserving of special commendation. Yet, the host communities also benefited from such improved properties, for under the state's so-called Aylmer Act, private properties within the Seashore's bounds were now subject to valuation and taxation by the towns. But the cultural intent of the authorizing legislation—preservation of the simple, traditional, man-land relationships on Cape Cod—now seems to have been overcome by the passage of time and events. Yet, there is evidence that what later transpired had been predicted earlier. At the end of the exhaustive zoning bylaw discussions, National Park Service regional director Lemuel Garrison was asked whether he wished the guidelines to be adopted as standards by secretarial regulation, including a section prescribing the character and degree of development permitted residences then in existence. He decided against such a procedure.[240]

In retrospect, it is interesting that this pioneering entry of the federal government into local land-use controls commanded so little congressional attention at the time. Far more significant issues were the requested contract authority for lands in advance of actual appropriation, a device not utilized until the Redwood National Park authorization a decade later, and the matter of payments-in-lieu-of-taxes. On the latter issue, the Seashore would have to await the generic legislation recommended by the Public Land Law Review Commission in 1976. Ultimately, the six host communities became the beneficiaries of a five-year transition period of in-lieu-of-tax payments, to be followed by a more modest payment of seventy-five cents per acre for what the federal government termed "entitlement" lands. This was at least a modest vindication of the long battle waged by Samuel Levy, chairman of Truro's Board of Selectmen, on the in-lieu-of-tax payment issue.[241]

THE ADVISORY COMMISSION

Once appointed to the Advisory Commission, the members made a point of attending meetings with great regularity. During the period 1962–81, one hundred and forty-four formal meetings were held. Attendance by voting members averaged an astonishing 79 percent. This spoke well both for the Commission and the nature of its business. While members obviously took their responsibilities seriously, they also found the meetings to be of considerable value and assistance. With the exception of a short period in the late 1960s, when the Commission and the Seashore were undergoing personnel and program changes, the rate of attendance has remained high. During the most recent five-year period, for example, attendance is not appreciably different than during the first five years.

A closer look at attendance statistics reveals no occasion in twenty years when less than half of the voting members were present.[242] On the rare occasions when absences occurred, they were generally for good reason. And even in those instances, the absent member invariably arranged in advance for an official observer to attend in his place.

Variability of attendance was extremely low for more than half of the jurisdictions. Eight out of the nine jurisdictions could claim at least one year when attendance was perfect. Only four jurisdictions displayed any real variation in annual attendance, and most of the variance is attributable to problems of extended illness or delays in the appointment of replacement members by the secretary of the interior. The near 90 percent record of attendance by the Barnstable County, Wellfleet, Eastham, and Chatham members is nothing short of extraordinary. Although partially explained by their status as officers or retirees, the simple truth is that each of these individuals became personally dedicated to the work of the Advisory Commission and to the future of the Seashore.

Conscientious attendance seemed to occur regardless of the jurisdiction represented. The federal member, for example, was not more diligent just because the Seashore was a federal undertaking. The state's attendance record appears appreciably lower—in part because the state had two places to account for on the Advisory Commission instead of just one.

Neither is any distinct attendance pattern discernible among the respective towns. One might suspect that those representatives whose towns had the least land would be the less frequent attendees, whereas representatives from towns more severely impacted by the Seashore would make a point of not missing a meeting. Yet the representative from the town with the least Seashore land (Chatham) had the best attendance record, and the representative from a town with considerable federal property (Truro) fell just below average in rate of attendance.

In 1969 the reduced volume of Seashore business and the pressing time demands of Commission meetings caused the Advisory Commission to move away from its original schedule of monthly meetings. Again, attendance rates remained high regardless of the number of meetings per year. One can only conclude that what the Commission does is infinitely more significant to attendance than how often it meets. If extent of participation is a key criterion, then the Advisory Commission still has a vital role to play in the affairs of the Seashore.

None would deny that the Advisory Commission has been an active participant in the development of the Seashore. But has it really been a success? The answer would seem to be yes as measured by most normal standards. Its influence on Seashore policy has been significant. It has been encouraged and enabled to carry out its assignment by the National Park Service. And the results of its efforts have been perceived to be useful by all parties concerned.

In terms of policy, the Advisory Commission has proved influential in a host of operational and developmental decisions. It has spoken strongly and effectively in favor of conservation as the first objective of the Seashore, even in the face of a national movement toward outdoor recreation. The issue was joined at the highest levels in Washington and, ultimately, in the language of an appropriations bill before Congress.

The Advisory Commission examined and modified two versions of the master plan for the Seashore, both to ensure its conservation principles and to reshape its development projects. The Advisory Commission also had much to do with the building of a local constituency for the Seashore from its earliest involvement in the preparation of zoning bylaws to the encouragement of cooperative working relations with the surrounding communities in such areas as police and fire protection, water supply, pest control, and education. For the most part, the Commission has echoed the expectations of the Senate Interior Committee which called for action without "a direct, personal interest" or "a strictly parochial attitude."[243]

Yet a good measure of the Commission's success is attributable to the attitude of the National Park Service. Despite earlier objections to the Advisory Commission, and the watering down of its proposed authority, tenure, and manner of operation, the Park Service has, in fact, consulted regularly with the Advisory Commission before making any major decision relative to the Seashore, just as Congress intended.

There is also abundant evidence that the Commission is viewed as a genuine asset by all participants. For example, the responses from the commonwealth, county, and towns to Chairman Nickerson's inquiry of February 27, 1975, described the Advisory Commission as a "bridge,"[244] "the only vehicle available to participate in policy formulation,"[245] "a device to maintain lines of communication,"[246] a "valuable service,"[247] and "in the best interests of the National Seashore Park and the Lower Cape communities."[248]

Superintendent Arnberger's internal memorandum to Associate Director Hulett, via the Northeast Regional Office, dated October 27, 1972, cited the Commission's valuable assistance in such matters as local and community affairs, marine sanctuary protection, and legislative authorization and funds. Arnberger recommended the Commission's continuation beyond its statutory termination date, a position that has been supported by every other superintendent with Cape Cod experience and both past and present regional office personnel.

Robert M. Landau, the deputy associate director of the Park Service responsible for advisory commission liaison, wary of the mixed quality of such bodies, made a point of attending the May 1975 meeting where a heated discussion took place over possible extension of the Provincetown Airport. Landau returned to Washington convinced that regional advisory bodies could not deal effectively with specific park business and sufficiently impressed to pledge his personal support for the continuation of the Cape Cod Advisory Commission.[249]

At the congressional hearings in 1970 on the expansion of the Seashore's authorization to $33.5 million, observers reported that Senate and House committee members were aware and appreciative of the real assistance being provided by the Advisory Commission.[250] One could almost hear Eastham's Harry Taylor, speaking at the Senate hearings on Cape Cod in December of 1959. The then-proposed Commission had indeed become "the last voice of the people."[251]

But if the Cape Cod Advisory Commission really did work, *why* had it worked so well? Several factors appear to have been responsible.

Whether by accident or design, the membership of the Advisory Commission seemed well fitted to the mission of the Seashore. Ten appeared to be just the right number of members—large enough for proper representation, yet not too large for constructive collaboration. The potential parochialism of six town nominees was counterbalanced by four members with broader responsibilities: two from the state and one each from the federal government and the county. The addition of a Barnstable County representative following testimony at the legislative hearings proved to be particularly useful, not only for the caliber of its particular representative (Joshua A. Nickerson) but for the representation afforded towns outside the Seashore's boundaries but impacted by its visitation.

As it turned out, the nominations of the local jurisdictions became the official Advisory Commission appointments. The secretary of the interior was well advised to let these nominations stand, despite the inclusion of a substantial number of individuals who had been working actively against the Seashore. The Advisory Commission was thus assured of a full spectrum of viewpoints and a membership, for the most part, with official and often elective responsibilities. Unlike many national advisory bodies, the Cape Cod Advisory Commissioners had a ring of reality to them. They had roots in the Seashore communities, and they cared about what it might become.

From the outset, the Advisory Commission had a clear sense of purpose and direction. It had strong chairmen and was willing to accept their leadership. A spirit of independence pervaded the Commission, nurtured by the legislative proceedings and the inadvertent first meeting in Washington. The National Park Service wisely acceded to and even encouraged a forthright expression of views, for militant independence had an obverse side that was an absolute necessity for the Seashore—the willingness of key opinion leaders to become directly involved in the shaping of its programs and policies.

The attitude of the National Park Service toward the Advisory Commission certainly had much to do with the Commission's success. It is difficult to account for the abrupt transition by the Service from the hard-nosed "we know best" position expressed to Congress to the fully cooperative attitude exhibited in later years. Fortunately for the Seashore, Park Service policy appears to have bowed to the realities of Park Service practice. A superb choice of Seashore superintendents, a happy accident of regional administrative personnel who were particularly sensitive to public participation considerations, and an unusual freedom to experiment in the face of a landmark seashore, all contributed to a thriving environment for a well-established local Advisory Commission.

A final factor was the particular public climate at the time of the creation of the Advisory Commission. Many people were growing weary of the almost two years of acrimony and conflict over the proposed Seashore. The issues were not so much whether the conservation of the lower Cape should take place, but how it should be accomplished. Now that the legislative issue had been settled, the Advisory Commission provided just that opportunity. The pragmatic Cape Codders recognized that there was going to be a Seashore. It was, therefore, in their interests to play an effective role in its implementation. For most of the Advisory Commission members, the time was ripe for collaboration.

The Cape Cod Advisory Commission engaged in an unusual amount of self-scrutiny during its first decade of existence due in part to three factors. First was the succession of superintendents—five in twenty years' time. It was only natural for the Commission to examine itself in preparing for the arrival of each new superintendent.

A second set of opportunities grew out of the Cape Cod Advisory Commission's emerging national reputation. Both on its own merits, and as a pacesetter for the National Park system as a whole, Cape Cod was to be consulted and usually visited by most of the new advisory bodies created by park enabling acts. Admirers within the National Park Service, such as Allen Edmunds, the planner responsible for the original Seashore survey report but later reassigned to the Midwest Regional Office, encouraged commissions from Indiana Dunes, Sleeping Bear, and the Ozark National Scenic Riverway to confer directly with their Cape Cod colleagues, but the roster also included the Fire Island National Seashore, the proposed Connecticut River Gateway National Recreation Area, the Padre Island National Seashore, the Assateague Island National Seashore, the Golden Gate National Recreation Area, the Delaware River Basin Commission, and several foreign delegations. Outside observers would normally include the superintendent, a leading member of his advisory body, and an official of an interest group concerned with the park. Occasions ranged from the private jet excursion of the Indiana Dunes Advisory Commission,[252] to the thoughtful discussions with the Minuteman Advisory Commission overlooking the "rude bridge" in Concord, Massachusetts.[253] Invariably, the highlight of each visit would be attendance at a meeting of the celebrated Cape Cod Commission.

A third reason for self-examination was one of necessity. The statutory life of the Cape Cod Advisory Commission would expire in May of 1976. Passage of the Federal Advisory Committee Act in 1972 virtually guaranteed that the federal government would take a far less liberal view of such advisory bodies. Both the congressional and the executive branches were moving to streamline the government advisory process, and Cape Cod could well be among the first fatalities.

Additional perspectives on the Advisory Commission experience can be drawn from three sources: the superintendents who served with the Advisory Commission, the Park Service officials knowledgeable of the Commission but detached from day-to-day contact, and the members of the Advisory Commission itself. If indeed Cape Cod serves as a valuable benchmark of effective public participation, a closer examination of these impressions would seem warranted.

Former superintendent Robert F. Gibbs was the first to be exposed to the Advisory Commission. Gibbs had no prior experience with park advisory bodies, nor was he familiar with the New England town meeting philosophy and style of business. Despite the unfortunate first meeting in Washington which he attended, he came away with the impression that the Advisory Commission members were a sophisticated, sincere group dedicated to the success of the project. His policies were to be as frank and cooperative as possible and to see that the Seashore earned the full respect of the Advisory Commission and the Cape Cod community. To this day, Gibbs feels that the Advisory Commission was an excellent idea and one with great advantage to all participants. He would favor the establishment of such bodies for every major area in the national park system.[254]

This policy of candor was carried forward by Gibbs's successor, Superintendent Stanley C. Joseph, who regarded the Advisory Commission as a key factor in the development of the Seashore. Among qualities producing well-reasoned advice, Joseph cited the Commission's diverse views, its combination of well-qualified people and excellent leadership, and its willingness to act as a group once a decision had been reached.[255]

The third superintendent, Leslie P. Arnberger, has observed that the Cape Cod Commission is probably one of the reasons the National Park Service has moved to establish similar bodies within each of its regions. Through the quality and motivation of its members and the excellence of its leadership, the Advisory Commission has been able to validate its actions in the public mind by deliberate policies of community involvement and openness right from the beginning.[256] In correspondence dated April 28, 1971, with the new executive director of the Midwest Regional Advisory Board, then-superintendent Arnberger laid down six basic principles for utilizing an Advisory Commission: keep it fully informed (even beyond statutory requirements), encourage staunch independence, keep Park Service presentations objective, be open and candid, develop mutual trust, and demand competent, intelligent, and objective leadership at all times.

The fourth superintendent, Lawrence C. Hadley, a veteran of nearly thirty years in the National Park Service and a native New Englander, has commended the sense of confidence and trust that exists between the Advisory Commission and the Seashore, adding that its effectiveness is well illustrated by the high public and community perception of its value. Because the Seashore is so omnipotently involved in the life-style and pattern of living on the lower Cape, the Commission's participation in the affairs of the Seashore is simply invaluable. The institution should be continued much as it is today, Hadley observed.[257]

For current superintendent Herbert Olsen, with his Scandinavian affection for the sea, the Cape Cod National Seashore has been a welcome assignment. Fresh from the Gateway National Recreation Area and prior service at the Minuteman National Historical Park, where the delivery of citizen advice never seemed to command a genuine priority, Olsen found the Advisory Commission's active interest a refreshing change. Even twenty years after its establishment, the Advisory Commission continues to meet bimonthly, its agendas crowded with current business and its members invariably present and prepared for thoughtful discussions. The number and scale of the receiving communities are what has made the difference, Olsen observed, for in a region as crowded as the lower Cape, it is imperative that effective interaction take place between the Seashore and its host communities.[258]

Former northeast regional director Ronald F. Lee in his retirement remarks in 1966 spoke of the fruitful and meaningful collaboration achieved by the Advisory Commission.[259] Lee made two significant observations: the complex, modern problems of conservation which, in an area like Cape Cod, virtually demand the broad-based advice of such a body; and the likelihood that this advice will be needed as much in the future as in the past. How prophetic those observations have proved to be, for the Advisory Commission today is busier than ever. Yet the agenda has changed appreciably from the early days of planning, development, and land acquisition concerns within the established boundaries of the Seashore. Areawide issues such as water supply, waste disposal, and mosquito control are now apt to appear on the Commission's docket. The most intractable of management problems (e.g., ORV usage) are the ones remaining. The weight of past discussions has been heavy on conservation. Because of growing visitor use, future policy debates may have to revolve around the accommodation of recreation facility demands. And the Seashore and its Advisory Commission will almost certainly need to reach out more to help the lower Cape communities cope with what Truro's Dexter Keezer has termed the "backwash" of the Seashore's conservation successes,[260] the resultant pressures for development in the privately-owned portions of the six Seashore towns.

The Commission's first chairman, Charles H. W. Foster, has observed that the Advisory Commission became a working commission the very day of the unfortunate first meeting in Washington. A mutual sense of outrage forged a curious alliance between proponents and opponents, a militant sense of independence, and an unspoken agreement to take the best of the viewpoints and make the experiment work. It is also possible that the depth of caring and concern expressed that day was not lost on the National Park Service, particularly the local and regional staff most directly involved in the implementation of the legislation. From that day forth, Foster has observed, the Seashore met its Advisory Commission more than halfway.[261]

Foster's successor as chairman, Joshua A. Nickerson, in 1971 correspondence with Chairman Carl T. Johnson of the Sleeping Bear Dunes National Lakeshore Advisory Commission, listed as the Cape Cod Commission's primary objectives the provision of local background information for the National Park Service, the protection of both park and community, the interpretation of park policies to the people (and vice versa), and the forceful presentation of suggestions "within the family."

Nickerson's successors have not disagreed, but their leadership has been more muted. Soft-spoken, low-key David Ryder, a fishing captain for his entire working life, found the Advisory Commission less controversial than he thought it would be. As chairman, it was his practice to ensure a dignified and orderly meeting and to seek consensus where possible.[262] The current chairman, Clifford White, a computer management expert by day and an ardent saltwater sports fisherman by night, has commended the Advisory Commission's role as a buffer between the area communities and the Seashore. The clear-cut definition of membership, the healthy mix of representation, and the high caliber of individuals chosen to serve have all contributed to an institution with credibility and influence, in his opinion.[263]

From the viewpoint of its members, the Advisory Commission also received generally high marks.[264] Most found the Commission a useful way of reaching a big government bureaucracy. But the reverse was also true. The Commission served as a stabilizing and mobilizing device for the collaboration of six highly independent Cape Cod communities. Far from just a sop to local interests thrown in at the time of the enabling legislation, the Advisory Commission became a valuable safety valve for Seashore decisions that could otherwise have become controversial.

Some members were apprehensive at first, fearing either meetings without substance or, worse yet, equivalents of the rancorous local selectmen's meetings. Still others assumed the Commission would be a closed corporation and were surprised by the openness of the discussions and the breadth of the agenda material. Curiously, there were no voting blocs and no lobbying for votes, which were usually not close even when taken. In its role as liaison between the management and the citizenry, the Advisory Commission seemed to serve equally often as sounding board and "fall guy" for the Seashore. And more than one Commission member has spoken warmly of what the experience did for the individual: the educational exposure to issues, the necessity to apply science in a management context, the reconciliation of varying and even opposing viewpoints, and the rewards of simple human collegiality. The Cape Cod National Seashore Advisory Commission *is* materially different than other advisory bodies, its members stoutly maintain, by virtue of its clear-cut membership and mandate, and the territorial imperative that encourages regular attendance by local representatives. Over the years, a general rapport with the Seashore began to emerge. As one member observed, the Advisory Commission, for the most part, became a "damned cohesive group" without the slightest sense of subservience to the Seashore administration. And because of its representative quality, it was a group to which the public could generally relate.

Perhaps special mention should be made of a particular representative quality. A measure of "loyal opposition" was assured until the involuntary retirement of Wellfleet representative Esther Wiles in August of 1974. Suspicious, conservative, and invariably outspoken, Mrs. Wiles was the epitome of Yankee self-government which, in Superintendent Hadley's words, "assured consistent direction and led to the acceptance and maturing of the Seashore."[265] To the end of her term, she remained steadfast in the belief that the town of Wellfleet was "too small an area for the amount of park we have,"[266] and never forgave the National Park Service for going back on her recollection of Director Conrad Wirth's public statement that a person "could do anything after the park as before."[267] Burr under the saddle that she was, Mrs. Wiles's departure was a genuine disappointment to the Commission for, as one member observed, she made Commission members "do a little thinking about both sides of a controversy."[268]

Yet, her response to Secretary of the Interior Roger Morton's recognition of her long service, written in January of 1975, had characteristic dignity. "It has been a privilege and a pleasure to be able to serve my country," Mrs. Wiles wrote. "In trying to uphold the principles of justice, I have been aware that the greatness of a nation is the sum total of the integrity of its people." None could have said it better.[269]

In later years, the Advisory Commission had begun to take on the characteristics of a "good old boys' club."[270] Once appointed, the members tended to remain for successive terms. The absence of a spectrum of ages, sexes, and occupations was noticeable. There was some talk that the Advisory Commission, in Esther Wiles's terms, had indeed become a bunch of "yes-men." Yet, appearances could be deceiving. For example, the presence of a contingent of veteran members assured an invaluable institutional memory, particularly when Seashore personnel were subject to transfer at two- to four-year intervals. Older members could be expected to make the significant commitments of time and interest required. They invariably came prepared. The Cape Cod retirement community had turned out to be an especially rich lode of management competence and experience to draw upon. Moreover, the local nominations process provided an opportunity for change every two years. It also guaranteed variability in membership. If the Advisory Commission representative was out of touch with his community, the fault lay squarely at the doorstep of the local selectmen who made the nomination. And what was needed next had already begun to happen. A citizen constituency, the Association for the Preservation of Cape Cod, had joined the lower Cape press in attending meetings and monitoring performance on a regular basis. In the future, the Association could be expected to lobby for quality nominations in the interest of ensuring an alert, representative, and responsive Advisory Commission.

Despite the general level of approbation, the Advisory Commission was still subject to improvement, many agreed. Over the years, a number of suggestions had emerged.

The Commission had been criticized with some justification for its preoccupation with detail. But, as one member observed,[271] a strength of the Commission was its practice of testing out policy decisions through careful scrutiny of their later application. If it were to relate only to policy, the Cape Cod Commission might lose much of the intimacy that characterized its relationship with the Seashore and underlay its credibility with the public.

The relationship of Commission members to their respective communities had never been entirely clear. Some towns chose selectmen to represent them on the Advisory Commission. Other communities selected prominent local citizens either because their elected officials were too busy to serve or because the officials preferred to stand a step removed from Seashore decisions.

The method of appointment had also received some criticism. Why should not the local jurisdictions name their representatives directly, rather than the present formality of having the secretary of the interior make the actual appointments? Although the secretary had never yet failed to respect a local nominee, there was the inference of a hand-picked commission. Counter to this argument was the added stature and prestige of members serving as United States advisory commissioners, which encouraged a broader view of their official responsibilities. At the least, however, should not the Advisory Commission be empowered to elect its own chairman? The new administrative charter in 1976 concurred with that viewpoint.

The composition of the Advisory Commission had also come in for critical examination. While all were agreed that the six town nominees were desirable, less certainty existed about the county, commonwealth, and Interior designees. Perhaps one state official would suffice, and perhaps the next county designee should come from the Cape Cod Planning and Economic Development Commission, not the Barnstable County Commissioners. Harking back to Charles Eliot's testimony at the Senate hearings in Eastham,[272] perhaps several of the Commission members should be required to have scientific or professional credentials. Over time, however, such members had naturally been appointed. Modifications were better enabled than prescribed, most observers agreed.

Another area of discussion involved the specific powers of the Advisory Commission. There had been much local testimony at the time of the hearings favoring a required consent of the Advisory Commission for certain Seashore actions. But the experience of ten years' time had confirmed the wisdom of former U.S. attorney general Francis Biddle's recommendation that the Advisory Commission be purely advisory.[273] Far from being constrained by lack of authority, the Commission had actually gained latitude in expressing its opinions, unencumbered as it now was by any official responsibility for its recommendations. However, experience had shown that the statutory consultations for commercial and recreational developments should have been broadened to include a direct responsibility for reviewing the Seashore's master plans.

Yet, for Cape Cod, these suggestions were almost moot, for neither the superintendent nor the Advisory Commission had ever hesitated in raising a matter for review. The policy from the very beginning had been one of consultation on every significant policy or program issue. The provisions section 8(f) of the Seashore Act ("shall, from time to time, consult . . . with respect to matters relating to the development of Cape Cod National Seashore") had been taken in their broadest context.

To sum up, the Advisory Commission, in the judgment of one veteran observer,[274] might have proved to be "a better thing for local folk than the local folk felt it would be," but it had clearly demonstrated its value. Conrad Wirth's observation that advisory bodies "start out good and then go sour on you" had not borne fruit on Cape Cod.[275] The device that had never been favored by the National Park Service, the Bureau of the Budget, and even the Senate Interior Committee staff, and by all rules should not have worked at all, had confounded the experts. As the Senate Committee on Interior and Insular Affairs had observed nearly fifteen years earlier, the role of the Advisory Commission could not be specified in rigid detail. It should properly be left to the "wisdom and good judgment of the persons who will serve on the Commission and the responsible officials of the Department of the Interior whose joint task and opportunity it will be soundly to develop and administer the seashore."[276] They had done just that.

THE BIOREGIONAL EXPERIENCE

But what can be said about Cape Cod's experience in managing a significant bioregion? A number of points must be made at the outset.

1. *Until the creation of the Seashore, no single entity was responsible for the conservation of this major bioregion.* The record of the Seashore in carrying out its statutory responsibilities obscures its more fundamental role – that of implementer of a sound program of conservation for the Seashore region. In many ways, it has been the Advisory Commission that has kept successive park administrations' feet to the fire on this issue. Witness the long struggle for enunciation of the supremacy of the conservation objective over the recreation mission, a project that engaged the Advisory Commission's attention for more than a decade. Conservation was also manifest in the master plan discussions and many of the development decisions – notably those affecting the restoration of the Eastham beaches. The principle of consistency with conservation cropped up in many other advisory actions of the Commission, such as the safety improvements planned for the Provincetown Airport, the use of oversand vehicles in the Province Lands, and the issuance of commercial permits for motels in Eastham and Orleans. While it is not suggested that the National Park Service has been

adverse to a dominant conservation objective, the Advisory Commission's strong advocacy position helped the agency maintain a balance in an era when national outdoor recreation was of unusual prominence.

2. *Pressing Seashore problems literally mandated their consideration on a bioregional scale.* The Seashore and its Advisory Commission confronted many issues that proved to be transboundary in nature. As early as 1966, for example, the Advisory Commission undertook to examine the root causes of erosion in the Truro and Eastham portions of the outer beach. Scientists from Woods Hole confirmed the coastal dynamics causing erosion in the middle sections and accretion at either end of the thirty-nine mile stretch. It was deemed inevitable that the developed facilities at Coast Guard Beach would be washed away, and they were designed accordingly. The ponds management issue opened up jurisdictional questions far beyond the matter of recreational usage in the Truro and Wellfleet portions of the Seashore. Similarly, a simple request for an emergency well to serve Provincetown triggered eventual consideration of Cape-wide groundwater problems. The aquifer at Provincetown was clearly interlinked with those elsewhere. And on the relatively straightforward matter of Wellfleet's proposed new sanitary landfill, actually outside the Seashore's formal jurisdiction, the issue would later expand to require complex studies of the entire marsh ecosystem.

3. *Future Seashore issues are even more likely to be bioregional in character.* Already on the horizon is the controversial matter of mosquito control through water management and selective larviciding, long practiced on the Cape but subject to increasing scrutiny for their possible effects on aquatic vegetation and groundwater quality. Tufts University biologist Norton Nickerson, in the curious position of a former Advisory Commission member, former commissioner of the Cape Cod Mosquito Control Commission, and staunch wetland advocate, has remarked that few people today appreciate how unbearable the mosquito problem was before control measures were instituted.[277] Another regional issue, currently on the Seashore's back burner but part of the master plan approved in 1970, is the development of a comprehensive system of hiking, bicycle, and possibly horseback trails. Conversion of the Cape's abandoned railroad right of way through the state Department of Environmental Management's proposed "rail trail" could bring a spur of the regional system to the very doorstep of the Seashore at the former Whitehead Brothers sand quarry in Provincetown. And it is only a matter of time before atmospheric pollution issues, perhaps the ultimate in bioregional topics, receive serious attention on Cape Cod. Barnstable County's extensive freshwater streams and ponds are already among the most severely impacted of the state's water resources due to the effects of acid precipitation.

4. The Seashore could do more to stimulate attention to bioregional problems throughout the lower Cape. Long preoccupied with its own management problems and respectful of the independence of its host communities, the Seashore has never fulfilled its genuine potential to serve as a conservation resource for the entire lower Cape region. It has a critical mass of manpower and competence to draw upon—also the special objectivity and stature of a federal agency. It has a working history of cooperative service in other fields—law enforcement, fire fighting, and education. The Seashore's effective interpretive programs, if targeted locally as well as nationally, could readily reinforce the ecological imperative throughout the entire lower Cape. In so doing, the Seashore would also help itself, for, as the State of the National Parks reports have indicated, the future threats to the system are more likely to come from exterior forces than from within its own boundaries.[278] There are signs that an expanded conservation service role for the Seashore with respect to its communities would not be unwelcome. A timely prospect is Truro's current interest in protecting the resources of the entire Pamet River valley, a project that includes portions of the Seashore's own jurisdiction.[279] Similar opportunities may well arise in Provincetown Harbor, Wellfleet's Herring River marshes, the ponds section of Truro and Wellfleet, and the productive ecosystems of Nauset Marsh and Pleasant Bay. The situation is not without precedent, for it was the Seashore's own Advisory Commission that charted the path nearly two decades ago with the creation of the first Scientific Advisory Committee composed entirely of outside specialists.

But if these bioregional initiatives are to come to pass, it will take an institutional presence to raise the possibility and see the program through to completion. Fortunately, there are two elements working in that direction. First, the Seashore administration has a natural and legislative proclivity to work bioregionally in order to fulfill its own conservation mandate. And second, it has been encouraged to do so at every turn of the road by an alert and involved Advisory Commission, the Seashore's primary bioregional institution. What can one conclude from Cape Cod's twenty-year experience with such an institution?

In the preface to this account, it was speculated that matters of size, composition, and continuity were important contributors to the viability of any bioregional institution. In Cape Cod's case, a continuing sense of representativeness, reinforced by the process of periodic nominations and appointments, has enabled the Advisory Commission to remain a credible body locally for more than twenty years. Also it was postulated earlier that some form of legitimacy is a prerequisite for institutional viability. The Commission's statutory and administrative connections with the Seashore have given it added credentials—national as well as local. A broad mandate to advise, rather than specific authority to consent, has ensured an open agenda which, in turn, has helped the Advisory Commission to remain involved and responsive. The continuing uncertainty over who the Commission's "client" really is—the Seashore or the lower Cape communities—has turned out to be an asset rather than a liability. It has given the Advisory Commission more real influence in both spheres than its charter would warrant. And the Commission's reputation and role have been enhanced by the professionalism it has displayed throughout its history. Issues have been examined factually, discussed openly, and resolved impartially. No wonder that when the Commission has spoken, its constituency has tended to listen carefully.

That bioregional institutions can be suspect was noted earlier. The Cape Cod experience bears out that observation. The Advisory Commission was viewed equally warily by the local communities and the National Park Service until it had proven its worth. It is still a fragile institution, clinging to life through a two-year, administrative charter, and utterly dependent upon its component jurisdictions for an influx of concerned and committed members. Yet, the legislative device, handcrafted to soften the blow of a federalized Seashore, has blossomed into a full-fledged instrument for improving the management of the entire bioregion. For an investment of a bowl of chowder per person bimonthly, at a support cost of one-quarter person-year and $3,500 annually, the Seashore has gained incalculable returns. Although one is forced to conclude that much of the Cape Cod success story is a matter of happy circumstance—the right people being in the right place at the right time—other bioregions should fare so well.

Chronological Summary

PREAUTHORIZATION PERIOD

1602 – Bartholomew Gosnold, exploration of Cape Cod Bay

1605 – Samuel de Champlain, exploration of Nauset Inlet

1614 – John Smith, mapping expedition of New England and Cape Cod

November 11, 1620 – Signing of the Mayflower Compact, Provincetown Harbor

1637 – Colonization of Cape Cod begins, settlement of Sandwich

1700–1800s – Growth of fishing, farming, salt industries on Cape Cod

1849–1855 – Visits to Cape Cod by Henry David Thoreau

1870 – Railroad service to Provincetown, beginning of tourist industry

1892 – Survey report on Cape Cod by Trustees of Public Reservations (Mass.)

1914 – Completion of the Cape Cod Canal

1919 – First eastern national park authorized, Acadia National Park (Maine)

1937 – First national seashore authorized, Cape Hatteras (North Carolina)

September 29, 1939 – Survey report, proposed Cape Cod National Seashore and Historic Parkway

1955 – Vanishing shoreline report published, Gulf and Atlantic coasts

November 1, 1956 – First public coverage of possible national seashore by *The Cape Codder*

April, 1957 – First Cape Cod legislation introduced by Congressmen Boland, O'Neill, and Philbin

1957–1958 – National Park Service field survey of Cape Cod (Drs. William Randall and William Vinal)

March 1959 – Publication of *Cape Cod National Seashore: A Proposal*

March 23–24, 1959 – Public meetings in Eastham and Chatham, Director Conrad Wirth

May 1959 – Omnibus seashore legislation introduced by Senator Neuberger

July 1959 – Massachusetts senators agree to draft legislation, staff visit Cape Cod

September 3, 1959 – Kennedy-Saltonstall-Keith legislation introduced

December 9–10, 1959 – Field hearings in Eastham, Senate Subcommittee on Public Lands

March 1960 – Seashore economic impact survey published by Economic Development Associates

June 20–21, 1960 – Hearings in Washington, Senate and House Interior and Insular Affairs Committees

December 16–17, 1960 – Field hearings in Eastham, House Subcommittee on National Parks

February 1961 – Revised Saltonstall-Smith-Keith legislation introduced

March 6–8, 1961 – Hearings in Washington, House Interior and Insular Affairs Committee

March 9, 1961 – Hearings in Washington, Senate Interior and Insular Affairs Committee

June 20, 1961 – Report No. 428, Senate Interior and Insular Affairs Committee

July 3, 1961 – Report No. 673, House Interior and Insular Affairs Committee

August 1, 1961 – Report No. 831, Senate-House Conference Committee

August 2, 1961 – Conference Committee recommendations accepted by Congress

August 7, 1961 – Public Law 87–126 signed into law by President John F. Kennedy

POSTAUTHORIZATION PERIOD

October 1961 – Land acquisition officer George H. Thompson opens Seashore office

January 1962 – Cape Cod National Seashore Advisory Commission appointments announced

February 16, 1962 – First meeting of Advisory Commission, Washington

March 1962 – Superintendent Robert F. Gibbs assigned to Cape Cod

April 24–27, 1962 – Meetings with towns on zoning bylaws

June 1962 – First option for land approved for purchase

July 1962 – Stage 1 master plan for Seashore discussed with Advisory Commission

August 1962 – Decontamination of Camp Wellfleet complete

September 1962 – First land taking approved to prevent development within Seashore

October 25–29, 1962 – Advisory Commission trip to Cape Hatteras and other national areas

January 1963 – First Advisory Commission action on commercial and industrial permits

March 1963 – Solicitation of national architects for Seashore design policies

April 1963 – New beach at Race Point under construction

August 1963 – Eastham, Orleans, Truro, and Chatham zoning bylaws approved by the secretary of interior.

September 1963 – Joint meeting with Minuteman National Historical Park Advisory Commission

November 1963 – Salt Pond and Nauset Knoll motels issue
December 1963 – Proposals to rename Seashore in memory of President Kennedy
March 1964 – Association for the Improvement of Medicine issue
October 1964 – Outermost House dedication
October 1964 – Campground development issue
February 1965 – Job Corps Conservation Center under construction
April 1965 – Designation of Seashore as recreational area becomes issue
December 1965 – Superintendent Stanley C. Joseph assigned to Cape Cod
March 1966 – Corps of Engineers study of Pleasant Bay
April 1966 – Establishment of Scientific Advisory Committee
May 30 1966 – Formal establishment of Seashore and dedication of Salt Pond Visitor Center

POSTESTABLISHMENT DECADE

June 1966 – Provincetown zoning bylaw approved by the secretary of the interior
June 1966 – Congressman Keith introduces legislation raising Seashore authorization
August 1966 – Highland Light golf course issue
October 1966 – Great Island access issue
December 1966 – Joshua A. Nickerson appointed chairman of the Advisory Commission
January 1967 – Offshore oil and gas development issue
March 1967 – Wellfleet zoning bylaw approved by the secretary of the interior
April 1967 – Province Lands bicycle trails and Camp Wellfleet developments approved
July 1967 – Institution of fees for Seashore beach usage
February 1968 – Land exchange with Nauset Regional School District
March 1968 – Revision of Seashore master plan begins
April 1968 – Closing of Seashore lands office
July 1968 – First discussion of regional planning needs
August 1968 – Superintendent Leslie P. Arnberger assigned to Cape Cod
October 1968 – Joint meeting with Indiana Dunes National Lakeshore Advisory Commission
November 1968 – 1960 Seashore economic survey updated by Philip B. Herr and Associates
March 1969 – Initiation of environmental education (NEED) program
May 1969 – Dedication of Province Lands Visitor Center
June 1969 – Closing of Job Corps Conservation Center
July 1969 – New beach at Marconi area, Camp Wellfleet, in operation
February 1970 – Senate and House hearings on Seashore authorization legislation
September 1970 – Public hearings on master plan revisions

September 1970 – Passage of state Ocean Sanctuary legislation

September 1970 – New land acquisition office opened, $33.5 million lands authorization

January 1971 – State legislation filed for Nauset Beach State Park

October 1971 – Cape Cod water supply issue

January 1972 – Legislation filed by Senator Edward M. Kennedy for Nantucket Islands Trust

September 8, 1972 – One hundredth meeting of Advisory Commission, retirement of Robert McNeece

October 1972 – First open meeting of Advisory Commission

November 1972 – Passage of Federal Advisory Committee Act

May 1973 – Cooperative Research Unit established at Seashore

September 1973 – Nude bathing issue first discussed by Advisory Commission

January 1974 – Superintendent Lawrence C. Hadley assigned to Cape Cod

July 1974 – State legislation enacted permitting taxation of occupancy interests

February 1975 – Regulations issued prohibiting nude bathing

May 1975 – Eastham beaches development issue

August 1975 – Civil liberties confrontation on nude bathing regulations

September 1975 – Studds-Kennedy legislation filed to extend life of Advisory Commission (not enacted)

January 1976 – Proposed reserve fund for acquisition by condemnation

January 1976 – Increased use of the Seashore by hang gliders

April 1976 – Discussion of Georges Bank offshore oil development issue

THE SECOND DECADE

October 8, 1976 – First meeting of Advisory Commission under administrative charter

October 1976 – Subcommittees on ponds management and improved property issues appointed

December 1976 – Subcommittee on insect pest control issue appointed

February 1977 – Brooke-Studds legislation filed to amend the Seashore Act (not enacted)

February 1977 – Subcommittee on oversand vehicle usage issue appointed

April 1977 – Certificates of suspension of condemnation for commercial properties reviewed

June 1977 – House passage of acquisitions backlog appropriation

June 1977 – Truro raises matter of tax loss reimbursement

August 1977 – Subcommittee report on improved properties issue discussed

August 1977 – Subcommittee on motel concessions issue appointed

October 1977 – Resignation of original member Ralph A. Chase of Eastham

December 1977 – National Park Service management categories abolished

December 1977 – Subcommittee report on parks management issue discussed

August 1978 – David F. Ryder elected chairman of the Advisory Commission

February 6-7, 1968 – Eastham beach facilities washed away, Provincetown dikes breached by storm

August 1978 – Gasoline leak threatened Provincetown's South Hollow wellfield, temporary well permit issued

August 1978 – Subcommittee report on motel concession renewals issue discussed

November 1978 – Superintendent Herbert Olsen assigned to Cape Cod

January 1979 – Atlantic Coastal Laboratory established at former Mitre site

March 1979 – Subcommittee on Eastham beach development issue appointed

May 1979 – Public meeting held on oversand vehicle usage

June 1979 – Navigation improvements proposed for Provincetown Airport

August 1979 – All day Advisory Commission field trip to Province Lands

October 1979 – Research reports on effects of oversand vehicle usage discussed

October 1979 – Subcommittee on land acquisition issue appointed

December 1979 – Subcommittee report on oversand vehicle usage issue discussed

March 1980 – Subcommittee report on land acquisition issue discussed

May 1980 – Three-member commission of retired judges appointed to expedite condemnation cases

July 1980 – Oversand management plan now subject to full environmental impact analysis

September 1980 – Subcommittee report on Eastham beach development issue discussed

January 1981 – Advisory Commission recommendations formulated on oversand vehicle management

April 1981 – Research report on dunes revegetation discussed

April 1981 – Wellfleet landfill issue discussed

June 1981 – Detailed discussion of water supply situation on Cape Cod and future of Provincetown's South Hollow wellfield

August 7, 1981 – Ceremonies commemorating twentieth anniversary of Seashore authorization

Individuals Interviewed

Victor Adams, selectman, Barnstable, Massachusetts

Leslie P. Arnberger, superintendent, Cape Cod National Seashore, Cape Cod, Massachusetts

Wallace Bailey, director, Wellfleet Bay Sanctuary, Massachusetts Audubon Society, Wellfleet, Massachusetts

Elmer V. Buschman, legal assistant, Cooperative Activities Division, National Park Service, Washington, D.C.

Ralph A. Chase, member, Advisory Commission, Eastham, Massachusetts

Serge Chermayeff, landowner, Wellfleet, Massachusetts

Josiah H. Child, member, Advisory Commission, Provincetown, Massachusetts

Bernard Collins, selectman, Eastham, Massachusetts

Norman Cook, executive director, Cape Cod Chamber of Commerce, Centerville, Massachusetts

Prescott Cummings, selectman, Eastham, Massachusetts

Leo E. Diehl, member, Advisory Commission, Washington, D.C.

John R. Dyer, selectman, Truro, Massachusetts

Allen Edmunds, park planner, National Park Service, Philadelphia, Pennsylvania

Frank Falacci, editor of Cape Cod news, Boston Globe, Hyannis, Massachusetts

Charles E. Frazier, selectman, Wellfleet, Massachusetts

Robert F. Gibbs, superintendent, Cape Cod National Seashore, Cape Cod, Massachusetts

Vernon Gilbert, chief of interpretive programs, Cape Cod National Seashore, Cape Cod, Massachusetts

Lawrence C. Hadley, superintendent, Cape Cod National Seashore, Cape Cod, Massachusetts

Alice Hiscock, chairman, Planning Board, Chatham, Massachusetts

Malcolm Hobbs, editor and publisher, The Cape Codder, Orleans, Massachusetts

Frederick Holborn, legislative assistant, Senator John F. Kennedy, Washington, D.C.

Dr. L. Thomas Hopkins, landowner, Truro, Massachusetts

Frederick Jewell, landowner and historian, Eastham, Massachusetts

Stanley C. Joseph, superintendent, Cape Cod National Seashore, Cape Cod, Massachusetts

Thomas Kane, town clerk and tax collector, Truro, Massachusetts

Mrs. Egon Kattwinkel, landowner, Eastham, Massachusetts

Dexter M. Keezer, member, Advisory Commission, Truro, Massachusetts

Hastings Keith, United States representative, Washington, D.C.

James Killian, chief of planning, Cape Cod National Seashore, Cape Cod, Massachusetts

Robert Landau, deputy associate director for legislation, National Park Service, Washington, D.C.

Fred G. LaPiana, selectman, Eastham, Massachusetts
Shirley M. Luikens, Advisory Boards and Commissions, National Park Service, Washington, D.C.
Dr. Sally H. Lunt, member, Advisory Commission, Weston, Massachusetts
Nathan Malchman, member, Advisory Commission, Provincetown, Massachusetts
David B. H. Martin, legislative assistant, Senator Leverett Saltonstall, Washington, D.C.
Dr. Barbara S. Mayo, member, Advisory Commission, Provincetown, Massachusetts
Charles A. Mayo, selectman, Provincetown, Massachusetts
Robert A. McNeece, member, Advisory Commission, Chatham, Massachusetts
Jonathan Moore, member, Advisory Commission, Cambridge, Massachusetts
Joshua A. Nickerson, member, Advisory Commission, Chatham, Massachusetts
Dr. Norton H. Nickerson, member, Advisory Commission, Medford, Massachusetts
Gaston L. Norgeot, member, Advisory Commission, Orleans, Massachusetts
Herbert Olsen, superintendent, Cape Cod National Seashore, Cape Cod, Massachusetts
George Palmer, associate regional director, National Park Service, Philadelphia, Pennsylvania
F. Cliff Pearce, architect, Cape Cod National Seashore, Cape Cod, Massachusetts
Dr. William Randall, University of Massachusetts, Amherst, Massachusetts
Charles R. Rinaldi, administrative officer, Cape Cod National Seashore, Cape Cod, Massachusetts
Robert Robes, executive director, Cape Cod Planning and Economic Development Commission, Barnstable, Massachusetts
Wilfred Rogers, selectman, Wellfleet, Massachusetts
David F. Ryder, member, Advisory Commission, Chatham, Massachusetts
Leverett Saltonstall, United States senator, Boston, Massachusetts
Francis W. Sargent, governor of Massachusetts, Boston, Massachusetts
Bernice Shears, selectman, Provincetown, Massachusetts
John C. Snow, selectman, Provincetown, Massachusetts
George H. Thompson, land acquisition officer, Cape Cod National Seashore, Cape Cod, Massachusetts
Ben H. Thompson, associate director, National Park Service, Washington, D.C.
Paul Todd, landowner, Truro, Massachusetts
Clifford H. White, member, Advisory Commission, Wrentham, Massachusetts
Esther Wiles, member, Advisory Commission, Wellfleet, Massachusetts
Conrad Wirth, director, National Park Service, Washington, D.C.
John Worthington, selectman, Truro, Massachusetts
Paul Younger, landowner, Eastham, Massachusetts

Cape Cod National Seashore Advisory Commission Meetings

1962

February 16, 1962	May 18, 1962	August 3, 1962
March 9, 1962	June 8, 1962	September 7, 1962
April 13, 1962	July 6, 1962	December 14, 1962

1963

January 11, 1963	May 24, 1963	September 27, 1963
February 15, 1963	June 21, 1963	November 1, 1963
March 8, 1963	July 19, 1963	December 6, 1963
April 26, 1963	August 23, 1963	

1964

January 10, 1964	June 5, 1964	October 30, 1964
March 6, 1964	July 10, 1964	December 4, 1964
April 17, 1964	August 14, 1964	
May 15, 1964	September 18, 1964	

1965

January 8, 1965	June 4, 1965	October 15, 1965
February 19, 1965	June 25, 1965	November 19, 1965
April 9, 1965	August 6, 1965	December 10, 1965
May 7, 1965	September 10, 1965	

1966

January 14, 1966	May 20, 1966	September 23, 1966
February 25, 1966	June 24, 1966	October 21, 1966
March 18, 1966	July 22, 1966	December 2, 1966
April 15, 1966	August 26, 1966	

1967

January 6, 1967	May 5, 1967	September 15, 1967
February 10, 1967	June 2, 1967	October 10, 1967
March 3, 1967	July 7, 1967	November 17, 1967
April 14, 1967	August 4, 1967	December 1, 1967

1968

January 19, 1968	May 10, 1968	September 27, 1968
February 9, 1968	June 14, 1968	October 25, 1968
March 1, 1968	July 12, 1968	November 15, 1968
April 12, 1968	August 2, 1968	December 20, 1968

1969

February 28, 1969	June 20, 1969	December 5, 1969
March 28, 1969	September 5, 1969	
May 2, 1969	October 17, 1969	

1970

February 6, 1970	May 15, 1970	September 11, 1970
April 17, 1970	June 26, 1970	November 6, 1970

1971

January 15, 1971	May 21, 1971	October 1, 1971
March 26, 1971	July 16, 1971	November 19, 1971

1972

January 7, 1972	June 23, 1972	December 1, 1972
February 25, 1972	September 8, 1972	
May 5, 1972	October 20, 1972	

1973

February 23, 1973	June 22, 1973	November 30, 1973
May 4, 1973	September 21, 1973	

1974

February 22, 1974	July 12, 1974	November 22, 1974
May 3, 1974	September 20, 1974	

1975

February 14, 1975	July 11, 1975	November 7, 1975
May 9, 1975	September 5, 1975	

1976

January 23, 1976	October 8, 1976	December 10, 1976
April 16, 1976		

1977

February 4, 1977	June 17, 1977	October 14, 1977
April 1, 1977	August 19, 1977	December 9, 1977

1978

August 25, 1978	November 3, 1978

1979

January 19, 1979	June 1, 1979	October 19, 1979
March 23, 1979	August 24, 1979	December 14, 1979

1980

March 7, 1980	July 11, 1980	November 14, 1980
May 2, 1980	September 12, 1980	

1981

January 16, 1981	June 12, 1981	August 7, 1981
April 24, 1981		

Cape Cod National Seashore Advisory Commission Members

SECRETARY OF THE INTERIOR

Leo E. Diehl   Dr. George M. Woodwell   Paul F. Nace, Jr.
Chester A. Robinson, Jr. #Dr. Sally H. Lunt   Jonathan Moore

COMMONWEALTH OF MASSACHUSETTS

* Charles H. W. Foster   Arthur W. Brownell   *Clifford H. White
Josiah H. Child   Dr. Norton H. Nickerson
Robert L. Yasi   #Dr. Barbara S. Mayo

BARNSTABLE COUNTY

*# Joshua A. Nickerson

PROVINCETOWN

Nathan Malchman

TRURO

John R. Dyer, Jr.   John W. Carleton   Dexter M. Keezer
Harold J. Conklin   Stephen R. Perry

WELLFLEET

Esther Wiles   John H. Whorf   Francis R. King

EASTHAM

Ralph A. Chase   Elizabeth Worthing   Bernard Richardson

ORLEANS

Arthur Finlay   †Linnell E. Studley   Sherrill B. Smith, Jr.
Gaston L. Norgeot   Edward J. Smith

CHATHAM

† Robert A. McNeece   *David F. Ryder   Thomas R. Pennypacker II

* Chairman
# Vice-Chairman
† Secretary

APPENDIX E

Attendance by Voting Members, 1962–1981

JURISDICTION	1962	1963	1964	1965	1966	1967	1968	1969	1970	1971
Interior	9	10	9	10	11	11	9	5	4	5
Massachusetts*	15	17	16	19	16	9	8	3	4	7
Barnstable Co.	8	11	9	11	10	12	9	6	6	5
Provincetown	6	8	8	7	11	5	7	6	5	6
Truro	10	10	10	9	2	11	9	4	4	1
Wellfleet	10	9	10	11	10	12	11	6	5	5
Eastham	9	8	10	9	11	9	10	7	4	4
Orleans	8	7	5	3	6	2	6	1	3	4
Chatham	10	11	9	11	11	12	10	7	6	6
ADVISORY COMMISSION AS A WHOLE										
Attendance	85	91	86	90	88	83	79	45	41	43
Meetings	10	11	10	11	11	12	11	7	6	6
Percentage	85	83	86	82	80	69	72	64	68	72

* Massachusetts has two representatives; all other jurisdictions only one.

SOURCE: Minutes of Advisory Commission meetings, Cape Cod National Seashore.

1972	1973	1974	1975	1976	1977	1978	1979	1980	1981	TOTAL	PERCENTAGE
5	4	4	4	3	0	0	5	4	2	114	79
8	5	7	7	8	11	4	12	9	7	192	67
7	5	5	5	4	6	1	6	4	3	133	92
7	5	4	5	4	6	1	6	5	4	116	81
5	3	4	2	3	5	2	6	5	4	109	76
5	5	5	5	4	6	2	6	4	4	135	94
7	5	5	4	2	1	2	6	5	4	122	85
6	3	4	4	1	4	1	4	2	3	77	53
6	4	4	5	3	6	2	6	5	4	138	96
56	39	42	41	32	45	15	57	43	35	1,137	—
7	5	5	5	4	6	2	6	5	4	144	—
80	78	84	82	80	75	75	95	86	88	—	79

APPENDIX F

Administrative Charter—February 29, 1984

1. The official designation of the committee is the Cape Cod National Seashore Advisory Commission.

2. The purpose of the committee is to consult with the Secretary of the Interior, or his designee, with respect to matters relating to the development of the Cape Cod National Seashore, and with respect to carrying out the provisions of Sections 4 and 5 of the Act establishing the seashore. Additionally, the advice of the committee shall be sought before the issuance of permits for industrial or commercial use of property within the seashore, and before the establishment of any public use area for recreational activity within the seashore. The committee also renders advice on the establishment of regulations governing public uses within the seashore, and any proposed action which would create a substantive change in natural conditions or in existing structures within the seashore.

3. In view of the goals and purposes of the committee, it will be expected to continue beyond the foreseeable future. However, its continuation will be subject to biennial review and renewal as required by Section 14 of Public Law 92-463.

4. The committee reports to the Superintendent, Cape Cod National Seashore, South Wellfleet, Massachusetts 02663.

5. Support for the committee is provided by the National Park Service, Department of the Interior.

6. The duties of the committee are solely advisory and are as stated in paragraph 2 above. The committee renders advice by affirmative vote of a majority of its members.

7. The estimated annual operating cost of this committee is $3,500, which includes the cost of ¼ person-year of staff support.

8. The committee meets approximately six times a year.

9. The committee will terminate 2 years from the date this charter is filed, unless, prior to that date, renewal action is taken as set forth in paragraph 3 above.

10. The committee's membership was outlined in Public Law 87-126. Specified requirements of that Act were that the committee was to be composed of ten members, each appointed for a term of 2 years by the Secretary of the Interior as follows:

   a. Six members appointed from recommendations made by each of the boards of selectmen of the towns of Truro, Wellfleet, Eastham, Orleans, Chatham, and Provincetown;

   b. One member to be appointed from recommendations of the county commissioners of Barnstable County, Commonwealth of Massachusetts;

   c. Two members to be appointed from recommendations of the Governor of the Commonwealth of Massachusetts; and

   d. One member to be designated by the Secretary.

Previous experience in working with the committee indicates that the membership requirements of the statute provided for a well-balanced committee as now required by the Federal Advisory Committee Act. The number of members provided for adequate representation of varying interests and functions. Accordingly, the number of members will be continued at 10, and derived from sources set forth in Public Law 87-126. Appointments shall be for 2 years subject to the biennial review of the committee under the Federal Advisory Committee Act. Any vacancy in the committee shall be filled in the same manner in which the original appointment was made.

The chairman is elected by a majority vote of the members.

11. The committee is not composed of, but may include, formal subcommittees or subgroups and, in addition, there may be ad hoc committees formed for special purposes. Meetings of these groups are subject to the same requirements of the Federal Advisory Committee Act as are meetings of the full committee.

12. The committee was originally established by Section 8 of Public Law 87-126. The committee is necessary in connection with the performance of duties imposed on the Department by law, and its establishment is authorized by Public Law 91-383.

_____

(William Clark)
Secretary of the Interior

Date Signed: _____ 2-8-84 _____

Date Charter Filed: _____ 2-29-84 _____

Public Law 87-126
87th Congress, S. 857
August 7, 1961

An Act

To provide for the establishment of Cape Cod National Seashore.

Be it enacted by the Senate and House of Representatives of the United States of America in Congress assembled, That (a) the area comprising that portion of the land and waters located in the towns of Province-town, Truro, Wellfleet, Eastham, Orleans, and Chatham in the Com-monwealth of Massachusetts, and described in subsection (b), is des-ignated for establishment as Cape Cod National Seashore (hereinaf-ter referred to as "the seashore").

    Cape Cod National Seashore, Mass.

    (b) The area referred to in subsection (a) is described as follows:

    Beginning at a point in the Atlantic Ocean one-quarter of a mile due west of the mean low-water line of the Atlantic Ocean on Cape Cod at the westernmost extremity of Race Point, Province-town, Massachusetts;

    Estab-lishment

    thence from the point of beginning along a line a quarter of a mile offshore of and parallel to the mean low-water line of the Atlantic Ocean, Cape Cod Bay, and Provincetown Harbor in generally southerly, easterly, and northerly directions rounding Long Point and then southwesterly to a point a quarter of a mile off-shore of the mean low-water line on the harbor side of the dike depicted on the United States Geological Survey Provincetown quadrangle sheet (1949) crossing an arm of the Provincetown Harbor;

    thence northerly, along a line a quarter of a mile offshore of and parallel to the low-water line at the dike to a point easterly of the point of intersection of the said dike with the boundary of the Province Lands Reservation as depicted on the said Provincetown quadrangle sheet;

    thence westerly to the said point of intersection of the dike and the Province Lands Reservation boundary;

    thence along the boundaries of the Province Lands Reservation northwesterly, northeasterly, northerly, and easterly to the easternmost corner of the reserva-tion being near United States Route 6;

    thence leaving the said easternmost corner along an extension of the southerly reservation boundary line easterly to the northerly right-of-way line of United States Route 6;

    thence along the northerly right-of-way line of United States Route 6 in a general easterly direction crossing the Truro-Provincetown line and continuing in the town of Truro in a generally southeasterly direction to a point four-tenths of a mile southeasterly of the southerly right-of-way line of Highland Road;

thence easterly five-tenths of a mile to a point;

thence turning and running in a southeasterly direction paralleling the general alinement of United States Route 6 and generally distant therefrom five-tenths of a mile to a point approximately 700 feet northwesterly of Long Nook Road;

thence southwesterly along a ridge generally paralleling the alinement of Long Nook Road and distant approximately 700 feet therefrom to a point two-tenths of a mile northeasterly of the northerly right-of-way line of United States Route 6;

thence southeasterly paralleling the general alinement of United States Route 6 and generally distant two-tenths of a mile northeasterly thereof to a point 300 feet south of the southerly right-of-way line of Higgins Hollow Road;

thence in a general easterly direction paralleling the southerly alinement of Higgins Hollow Road and 300 feet distant southerly therefrom to a point five-tenths of a mile east of the easterly right-of-way line of said Route 6;

thence turning and running in a southeasterly and southerly direction paralleling the general alinement of United States Route 6 and distant five-tenths of a mile easterly therefrom to a point 300 feet north of the northerly right-of-way line of North Pamet Road;

thence in a generally southwesterly direction paralleling the general alinement of North Pamet Road and generally distant 300 feet northerly therefrom to a point approximately two-tenths of a mile east of the easterly right-of-way line of United States Route 6;

thence in a southerly direction paralleling the alinement of United States Route 6 and generally distant two-tenths of a mile easterly therefrom to a point three-tenths of a mile south of South Pamet Road;

thence west to the intersection of Old County Road and Mill Pond Road;

thence following the easterly right-of-way line of Old County Road southward to a point opposite the southerly right-of-way line of Ryder Beach Road at its intersection with Old County Road;

thence eastward to a point 300 feet east of the easterly right-of-way line of said Old County Road;

thence in a southerly direction paralleling Old County Road at a distance of 300 feet to the east of the easterly right-of-way line of said road to a point 600 feet south of the southerly right-of-way line of Prince Valley Road;

thence in a generally westerly direction, crossing Old County Road and the New York, New Haven, and Hartford Railroad right-of-way to the southern extremity of the town landing and beach in the Ryder Beach area, and continuing to a point in Cape Cod Bay a quarter of a mile offshore from the mean low-water line of Cape Cod Bay;

thence turning and running along a line a quarter of a mile offshore of and parallel to the mean low-water line of Cape Cod Bay in a general southerly and easterly direction rounding Jeremy Point and thence in a general northerly direction along a line a quarter of a mile offshore of and parallel to the mean low-water line on the westerly side of Wellfleet Harbor, to a point one quarter of a mile due north of the mean low-water line at the eastern tip of Great Island as depicted on the United States Geological Survey Wellfleet quadrangle sheet (1958);

thence north to the mean high-water line on the north shore of the Herring River estuary in the vicinity of its confluence with Wellfleet Harbor;

thence following the mean high-water line southwesterly, northwesterly, and northeasterly to the easterly right-of-way line of Chequesset Neck Road at its crossing of Herring River;

thence following the course of Herring River along the 20-foot contour line of the southeasterly shore thereof to a point near Mill Creek;

thence crossing Mill Creek in a northeasterly direction to the 20-foot contour level near to and northeast of the confluence of Mill Creek and Herring River;

thence following generally northerly and easterly along the easterly edge of the Herring River marshes on the 20-foot contour to a point north of which the easterly right-of-way line of a medium duty road, as depicted on said Wellfleet quadrangle sheet, crosses northward across a marshy stream near the juncture of said medium duty road with Bound Brook Island Road;

thence crossing said marshy stream along said easterly right-of-way line of said medium duty road, and continuing in a northerly direction to the 20-foot contour level on the north side of said marshy stream;

thence following the 20-foot contour line westward approximately 1,000 feet to its intersection with an unimproved dirt road, as depicted on said Wellfleet quadrangle sheet, leading from a point near the juncture of Bound Brook Island Road and the said medium duty road;

thence following said unimproved dirt road northwesterly for approximately 1,600 feet to the 20-foot contour line bordering the southerly edge of the Herring River marshes;

thence following said 20-foot contour line in an easterly direction to Route 6;

thence crossing Route 6 and continuing to a point on the easterly right-of-way line of a power transmission line as depicted on said Wellfleet quadrangle sheet;

thence in a general southerly direction along the said easterly right-of-way line of a power transmission line to the Eastham-Wellfleet town line;

thence southeasterly for a distance of approximately 5,200 feet to a point due north of the intersection of the easterly right-of-way line of Nauset Road with the northerly right-of-way line of Cable Road;

thence due south to the intersection of the said easterly right-of-way line of Nauset Road and the said northerly right-of-way line of Cable Road;

thence in a general southerly direction crossing Cable Road and along said easterly right-of-way line of Nauset Road to a point 500 feet north of the northerly right-of-way line of Doane Road and its intersection with Nauset Road;

thence west to a point 500 feet west of the westerly right-of-way line of Nauset Road;

thence southerly and westerly 500 feet from and parallel to the said right-of-way line of Nauset Road to the easterly right-of-way line of Salt Pond Road;

thence southerly along the easterly right-of-way line of said Salt Pond Road to its intersection with the southerly right-of-way line of Nauset Road;

thence westerly along the southerly right-of-way line of Nauset Road to its intersection with the easterly right-of-way line of United States Route 6;

thence southerly along the easterly right-of-way line of said Route 6 a distance of about four-tenths of a mile to the northerly boundary of the Eastham town hall property;

thence easterly to a point one-tenth of a mile from United States Route 6;

thence turning and running in a generally southerly direction paralleling the general alinement of United States Route 6 and generally distant therefrom one-tenth of a mile to a small stream approximately one-tenth of a mile beyond Governor Prence Road extended;

thence southeasterly along the said stream to the Orleans-Eastham town line;

thence along the Orleans-Eastham town line to the southerly tip of Stony Island;

thence generally southeasterly in the town of Orleans by Nauset Harbor Channel to a point due north of the northerly tip of Nauset Heights as depicted on United States Geological Survey Orleans quadrangle sheet (1946);

thence due south to the 20-foot contour line in Nauset Heights as delineated on the said Orleans quadrangle sheet;

thence generally southerly along the said 20-foot contour to a point about one-tenth of a mile northerly of Beach Road;

thence southwesterly along a line intersecting Beach Road at a point two-tenths of a mile easterly of the so-called Nauset Road leading northerly to Nauset Heights;

thence southerly to a head of a tributary to Little Pleasant Bay at the northerly tip of Pochet Neck as depicted on the said Orleans quadrangle sheet;

thence generally southerly along the thread of channel of the said tributary passing westerly and southwesterly around Pochet Island and thence southwesterly into Little Pleasant Bay passing to westerly of the northerly tip of Sampson Island, the westerly tip of Money Head, and the southwesterly tip of Hog Island following in general the centerline of Little Pleasant Bay to Pleasant Bay;

thence generally southeasterly in Pleasant Bay along a line passing midway between Sipson Island and Nauset Beach to a point on the Chatham-Orleans town line one-quarter of a mile westerly of the mean low-water line of Pleasant Bay on the westerly shore of Nauset Beach;

thence generally southerly in Pleasant Bay in the town of Chatham along a line a quarter of a mile offshore of and parallel to the said mean low-water line of Pleasant Bay on the westerly shore of Nauset Beach to a point a quarter of a mile south of the mean low-water line of the southern tip of Nauset Beach;

thence easterly rounding the southern tip of Nauset Beach along a line a quarter of a mile offshore of and parallel thereto;

thence generally northerly and northwesterly, and westerly along a line a quarter of a mile offshore of and parallel to the mean low-water line of the Atlantic Ocean on the easterly shore of Nauset Beach and on to the outer cape to the point of beginning.

SEC 2. (a) The Secretary of the Interior (hereinafter referred to as "Secretary") is authorized to acquire by purchase, gift, condemnation, transfer from any Federal agency, exchange, or otherwise, the land, waters, and other property, and improvements thereon and any interest therein, within the **Acquisition** area which is described in section 1 of this Act or which lies within **of land, etc.** the boundaries of the seashore as described pursuant to section 3 of this Act (both together hereinafter in this Act referred to as "such **Authority** area"). Any property, or interest therein, owned by the Commonwealth of Massachusetts, by any of the towns referred to in section 1 of this Act, or by any other political subdivision of said Commonwealth may be acquired only with the concurrence of such owner. Notwithstanding any other provision of law, any Federal property located within such area may, with the concurrence of the agency having custody thereof, be transferred without consideration to the administrative jurisdiction of the Secretary for use by him in carrying out the provisions of this Act.

(b) The Secretary is authorized (1) to use donated and appropriated funds in making acquisitions under this Act, and (2) to pay therefor not more than the fair market value of any acquisitions which he makes by pur- **Funds** chase under this Act.

(c) In exercising his authority to acquire property by exchange, the Secretary may accept title to any non-Federal property located within such area and convey to the grantor of such property any federally owned property under the jurisdiction of the Secretary within such area. The properties so exchanged shall be approximately equal in fair market value: *Provided,* That the Secretary may accept cash from or pay cash to the grantor in such an exchange in order to equalize the values of the properties exchanged.

The Secretary shall report to the Congress on every exchange carried out under authority of this Act within thirty days from its consummation, and each such report shall include a statement of the fair market values of the properties involved and of any cash equalization payment made or received. **Report to Congress**

(d) As used in this Act the term "fair market value" shall mean the fair market value as determined by the Secretary, who may in his discretion base his determination on an independent appraisal obtained by him. **"Fair market value"**

SEC 3. (a) As soon as practicable after the date of enactment of this Act and following the acquisition by the Secretary of an acreage in the area described in section 1 of this Act that is in the opinion of the Secretary efficiently administrable to carry out the purposes of this Act, the Secretary shall establish Cape Cod National Seashore by the publication of notice thereof in the Federal Register. **Notice** **Publication in F. R.**

(b) Such notice referred to in subsection (a) of this section shall contain a detailed description of the boundaries of the seashore which shall encompass an area as nearly as practicable identical to the area described in section 1 of this Act. The Secretary shall forthwith after the date of publication of such notice in the Federal Register (1) send a copy of such notice, together with a map showing such boundaries, by registered or certified mail to the Governor of the Commonwealth of Massachusetts and to the board of selectmen of each of the towns referred to in section 1 of this Act; (2) cause a copy of such notice and map to be published in one or more newspapers which circulate in each of such towns; and (3) cause a certified copy of such notice, a copy of such map, and a copy of this Act to be recorded at the registry of deeds for Barnstable County, Massachusetts.

SEC 4. (a) (1) The beneficial owner or owners, not being a corporation, of a freehold interest in improved property which the Secretary acquires by condemnation may elect, as a condition to such acquisition, to retain the right of use and occupancy of the said property for noncommercial residential purposes for a term of twenty-five years, or for such lesser time as the said owner or owners may elect at the time of such acquisition. **Acquisition by condemnation** **Provisions**

(2) The beneficial owner or owners, not being a corporation, of a freehold estate in improved property which property the Secretary acquires by condemnation, who held, on September 1, 1959, with respect to such property, an estate of the same nature and quality, may elect, as an alternative and not in addition to whatever right of election he or they might have under paragraph (1) of this subsection, to retain the right of use and occupancy of the said property for non-commercial residential purposes (i) for a term limited by the nature and quality of his or their said estate, if his or their said estate is a life estate or an estate pur auter vie, or (ii) for a term ending at the death of such owner or owners, or at the death of the survivor of them, if his or their said estate is an estate of fee simple.

(3) Where such property is held by a natural person or persons for his or their own life or lives or for the life or lives of another or others (such person or persons being hereinafter called "the life tenant"), with remainder in **"The life** another or others, any right of election provided for in paragraph (2) **tenant"** of this subsection shall be exercised by the life tenant, and any right of election provided for in paragraph (1) of this subsection shall be exercised by the concurrence of the life tenant and the remainderman or remaindermen.

(4) The beneficial owner or owners of a term of years in improved property which the Secretary acquires by condemnation may elect, as a condition to such acquisition, to retain the right of use and occupancy of the said property for noncommercial residential purposes for a term not to exceed the remainder of his or their said term of years, or a term of twenty-five years, whichever shall be the lesser. The owner or owners of the freehold estate or estates in such property may, subject to the right provided for in the preceding sentence, exercise such right or rights of election as remain to them under paragraphs (1) and (2) of this subsection.

(5) No right of election accorded by paragraphs (1), (2), or (4) of this subsection shall be exercised to impair substantially the interests of holders of encumbrances, liens, assessments, or other charges upon or against the property.

(6) Any right or rights of use and occupancy retained pursuant to paragraphs (1), (2), and (4) of this subsection shall be held to run with the land, and may be freely transferred and assigned.

(7) In any case where a right of use and occupancy for life or for a fixed term of years is retained as provided in paragraph (1), (2), or (4) of this subsection, the compensation paid by the Secretary for the property shall not exceed the fair market value of the property on the date of its acquisition by the Secretary, less the fair market value on such date of the said right retained.

(8) The Secretary shall have authority to terminate any right of use and occupancy of property, retained as provided in paragraph (1), (2), or (4) of this subsection, at any time after the date when any use occurs with respect to such property which fails to conform or is in any manner opposed to or inconsistent with any applicable standard contained in regulations issued pursuant to section 5 of this Act and in effect on said date: *Provided,* That no use which is in con- **Violation of** formity with the provisions of a zoning bylaw approved in accord- **regulations** ance with said section 5 which is in force and applicable to such property shall be held to fail to conform or be opposed to or inconsistent with any such standard. In the event that the Secretary exercises the authority conferred by this paragraph, he shall pay to the owner of the right so terminated an amount equal to the fair market value of the portion of said right which remained on the date of termination.

(b) (1) The Secretary's authority to acquire property by condemna- **Suspension** tion shall be suspended with respect to all improved property lo- **of authority** cated within such area in all of the towns referred to in section 1 of this Act for one year following the date of its enactment.

(2) Thereafter such authority shall be suspended with respect to all improved property located within such area in any one of such towns during all times when such town shall have in force and applicable to such property a duly adopted, valid zoning bylaw approved by the Secretary in accordance with the provisions of section 5 of this Act.

(c) The Secretary's authority to acquire property by condemnation shall be suspended with respect to any particular property which is used for commercial or industrial purposes during any periods when such use is permitted by the Secretary and during the pendency of the first application for such permission made to the Secretary after the date of enactment of this Act provided such application is made not later than the date of establishment of the seashore.

(d) The term "improved property," wherever used in this Act, shall mean a detached, one-family dwelling the construction of which was begun before September 1, 1959 (hereinafter referred to as "dwelling"), together with so much of the land on which the dwelling is situated, the said land being in the same ownership as the dwelling, as the Secretary shall designate to be reasonably necessary for the enjoyment of the dwelling for the sole purpose of noncom-   **"Improved** mercial residential use, together with any structures accessory to the   **property"** dwelling which are situated on the land so designated. The amount of the land so designated shall in every case be at least three acres in area, or all of such lesser amount as may be held in the same ownership as the dwelling, and in making such designation the Secretary shall take into account the manner of noncommercial residential use in which the dwelling and land have customarily been enjoyed: *Provided, however,* That the Secretary may exclude from the land so designated any beach or waters, together with so much of the land adjoining such beach or waters as the Secretary may deem necessary for public access thereto.

(e) Nothing in this section or elsewhere in this Act shall be construed to prohibit the use of condemnation as a means of acquiring a clear and marketable title, free of any and all encumbrances.

SEC 5. (a) As soon after the enactment of this Act as may be practicable, the Secretary shall issue regulations specifying standards for approval by him of zoning bylaws for purposes of section 4 of this Act. The Secretary   **Issuance of** may issue amended regulations specifying standards for approval by   **regulations** him of zoning bylaws whenever he shall consider such amended regulations to be desirable due to changed or unforeseen conditions.

All regulations and amended regulations proposed to be issued under authority of the two preceding sentences of this subsection shall be submitted to the Congress and to the towns named in section 1 of this Act at least ninety calen-   **Submission** dar days (which ninety days, however, shall not include days on   **to Congress** which either the House of Representatives or the Senate is not in session because of an adjournment of more than three calendar days to a day certain) before they become effective and the Secretary shall, before promulgating any such proposed regulations or amended regulations in final form, take due account of any suggestions for their modification which he may re-   **Publication** ceive during said ninety-day period. All such regulations and   **in F. R.** amended regulations shall, both in their proposed form and in their final form, be published in the Federal Register.

The Secretary shall approve any zoning bylaw and any amendment to any approved zoning bylaws submitted to him which conforms to the standards contained in the regulations in effect at the time of the adop-   **Zoning** tion by the town of such bylaw or such amendment unless before   **bylaws** the time of adoption he has submitted to the Congress and the towns and published in the Federal Register as aforesaid proposed amended   **Approval** regulations with which the bylaw or amendment would not be in conformity, in which case he may withhold his approval pending completion of the review and final publication provided for in this subsection and shall thereafter approve the bylaw or amendment only if it is in conformity with the amended regulations in their final form. Such approval shall not be withdrawn or revoked, nor shall its effect be altered for purposes of section 4 of this Act by issuance of any such amended regulations after the date of such approval, so long as such bylaw or such amendment remains in effect as approved.

(b) The standards specified in such regulations and amended regulations for approval of any zoning bylaw or zoning bylaw amendment shall contribute to the effect of (1) prohibiting the commercial and industrial use, other than any commercial or industrial use which is permitted by the Secretary, **Special** of all property within the boundaries of the seashore which is situ- **provisions** ated within the town adopting such bylaw; and (2) promoting the preservation and development, in accordance with the purposes of this Act, of the area comprising the seashore, by means of acreage, frontage, and setback requirements and other provisions which may be required by such regulations to be included in a zoning bylaw consistent with the laws of Massachusetts.

(c) No zoning bylaw or amendment of a zoning bylaw shall be approved by the Secretary which (1) contains any provision which he may consider adverse to the preservation and development, in accordance with the purposes of this Act, of the area comprising the seashore, or (2) fails to have the effect of providing that the Secretary shall receive notice of any variance granted under and any exception made to the application of such bylaw or amendment.

(d) If any improved property with respect to which the Secretary's authority to acquire by condemnation has been suspended by reason of the adoption and approval, in accordance with the foregoing provisions of this section, of a zoning bylaw applicable to such property (hereinafter referred to as "such bylaw")—

(1) is made the subject of a variance under or an exception to such bylaw, which variance or exception fails to conform or is in any manner opposed to or inconsistent with any applicable standard contained in the regulations issued pursuant to this section and in effect at the time of the passage of such bylaw, or

(2) is property upon or with respect to which there occurs any use, commencing after the date of the publication by the Secretary of such regulations, which fails to conform or is in any manner opposed to or inconsistent with any applicable standard contained in such regulations (but no use which is in conformity with the provisions of such bylaw shall be held to fail to conform or be opposed to or inconsistent with any such standard),

the Secretary may, at any time and in his discretion, terminate the suspension of his authority to acquire such improved property by condemnation: *Provided, however,* That the Secretary may agree with the owner or owners of such property to refrain from the exercise of the said authority during such time and upon such terms and conditions as the Secretary may deem to be in the best interests of the development and preservation of the seashore.

SEC 6. The Secretary shall furnish to any party in interest requesting the same, a certificate indicating, with respect to any property located within the seashore as to which the Secretary's authority to acquire such prop- **Certificate** erty by condemnation has been suspended in accordance with the provisions of this Act, that such authority has been so suspended and the reasons therefor.

SEC 7. (a) Except as otherwise provided in this Act, the property acquired by the Secretary under this Act shall be administered by the **Adminis-** Secretary subject to the provisions of the Act entitled "An Act to es- **tration** tablish a National Park Service, and for other purposes", approved

August 25, 1916 (39 Stat. 535), as amended and supplemented, and   **16 USC 1-4**
in accordance with laws of general application relating to the national
park system as defined by the Act of August 8, 1953 (67 Stat. 496);   **16 USC lb-ld**
except that authority otherwise available to the Secretary for the conservation and management of natural resources may be utilized to the extent he
finds such authority will further the purposes of this Act.

(b) (1) In order that the seashore shall be permanently preserved in its present
state, no development or plan for the convenience of visitors shall be undertaken
therein which would be incompatible with the preservation of the
unique flora and fauna or the physiographic conditions now prevail-   **Protection**
ing or with the preservation of such historic sites and structures as   **and**
the Secretary may designate: *Provided,* That the Secretary may pro-   **development**
vide for the public enjoyment and understanding of the unique natu-
ral, historic, and scientific features of Cape Cod within the seashore by establishing
such trails, observation points, and exhibits and providing such services as he may
deem desirable for such public enjoyment and understanding: *Provided further,* That
the Secretary may develop for appropriate public uses such portions of the seashore
as he deems especially adaptable for camping, swimming, boating, sailing, hunting,
fishing, the appreciation of historic sites and structures and natural features of Cape
Cod, and other activities of similar nature.

(2) In developing the seashore the Secretary shall provide public use areas in such
places and manner as he determines will not diminish for its owners or occupants
the value or enjoyment of any improved property located within the seashore.

(c) The Secretary may permit hunting and fishing, including shellfishing, on lands
and waters under his jurisdiction within the seashore in such areas and under such
regulations as he may prescribe during open seasons prescribed by
applicable local, State and Federal law. The Secretary shall consult   **Hunting and**
with officials of the Commonwealth of Massachusetts and any politi-   **fishing**
cal subdivision thereof who have jurisdiction of hunting and fishing,
including shellfishing, prior to the issuance of any such regulations,   **Regulations**
and the Secretary is authorized to enter into cooperative arrange-
ments with such officials regarding such hunting and fishing, including shellfishing, as he may deem desirable, except that the Secretary shall leave all aspects of the
propagation and taking of shellfish to the towns referred to in section 1 of this Act.

The Secretary shall not interfere with navigation of waters within
the boundaries of the Cape Cod National Seashore by such means   **Navigation**
and in such areas as is now customary.

SEC 8. (a) There is hereby established a Cape Cod National Sea-   **Cape Cod**
shore Advisory Commission (hereinafter referred to as the "Commis-   **National**
sion"). Said Commission shall terminate ten years after the date the   **Seashore**
seashore is established under section 3 of this Act.   **Advisory**
    Commission

(b) The Commission shall be composed of ten members each appointed for a term of two years by the Secretary as follows:

(1) Six members to be appointed from recommendations made   **Membership**
by each of the boards of selectmen of the towns referred to in the
first section of this Act, one member from the recommendations made by each
such board;

(2) One member to be appointed from recommendations of the county commissioners of Barnstable County, Commonwealth of Massachusetts;

(3) Two members to be appointed from recommendations of the Governor of
the Commonwealth of Massachusetts; and

(4) One member to be designated by the Secretary.

(c) The Secretary shall designate one member to be Chairman. Any vacancy in the Commission shall be filled in the same manner in which the original appointment was made.

(d) A member of the Commission shall serve without compensa- **Compensa-**
tion as such. The Secretary is authorized to pay the expenses reason- **tion**
ably incurred by the Commission in carrying out its responsibilities
under this Act upon vouchers signed by the Chairman.

(e) The Commission established by this section shall act and ad- **Duties**
vise by affirmative vote of a majority of the members thereof.

(f) The Secretary of his designee shall, from time to time, consult with the members of the Commission with respect to matters relating to the development of Cape Cod National Seashore and shall consult with the members with respect to carrying out the provisions of sections 4 and 5 of this Act.

(g) No permit for the commercial or industrial use of property located within the seashore shall be issued by the Secretary, nor shall any public use area for recreational activity be established by the Secretary within the seashore, without the advice of the Commission, if such advice is submitted within a reasonable time after it is sought.

(h) (1) Any member of the Advisory Commission appointed under **Exemptions**
this Act shall be exempted, with respect to such appointment, from
the operation of sections 281, 283, 284, and 1914 of title 18 of the **62 Stat. 697,**
United States Code and section 190 of the Revised Statutes (5 U.S.C. **793**
99) except as otherwise specified in subsection (2) of this section.

(2) The exemption granted by subsection (1) of this section shall not extend—

(i) to the receipt or payment of salary in connection with the appointee's Government service from any sources other than the private employer of the appointee at the time of his appointment; or

(ii) during the period of such appointment, and the further period of two years after the termination thereof, to the prosecution or participation in the prosecution, by any person so appointed, of any claim against the Government involving any matter concerning which the appointee had any responsibility arising out of his appointment during the period of such appointment.

SEC 9. There are authorized to be appropriated such sums as may be necessary to carry out the provisions of this Act; except that no more than
$16,000,000 shall be appropriated for the acquisition of land and **Appropria-**
waters and improvements thereon, and interests therein, and inci- **tion**
dental costs relating thereto, in accordance with the provisions of
this Act.

SEC 10. If any provision of this Act or the application of such provision to any person or circumstance is held invalid, the remainder of **Separability**
this Act or the application of such provision to persons or circumstances other than those to which it is held invalid shall not be affected thereby.

Approved August 7, 1961, 12:00 A.M.

Notes

1. Advisory Commission on Intergovernmental Relations, *Multistate Regionalism*, 1972.
2. *Enterprise and Journal* (Orange, Mass.), March 26, 1959.
3. Robert F. Gibbs, "Cape Cod National Seashore Case Study," November 1975 (unpublished); Kittredge, *Cape Cod*, [1930] 1968; Freeman, *History of Cape Cod*, 1862.
4. Beston, *The Outermost House*, [1928] 1983.
5. Charles H. W. Foster, "The New England Town Meeting," March 1968 (unpublished).
6. Blair Associates, "A Study of Eastham," submitted at the Eastham (Mass.) hearing, Senate Subcommittee on Public Lands, December 9, 1959.
7. Interview with Leverett Saltonstall, November 18, 1975.
8. *Enterprise and Journal* (Orange, Mass.), March 26, 1959.
9. Eastham (Mass.) hearings: Senate Subcommittee on Public Lands, December 9–10, 1959; House Subcommittee on Public Lands, December 16–17, 1960.
10. See Appendix G.
11. Statement made on the floor of the Senate by Senator John F. Kennedy, September 4, 1959.
12. Eastham (Mass.) hearing, Senate Subcommittee on Public Lands, December 9–10, 1959.
13. Senate Report No. 428, Senate Committee on Interior and Insular Affairs, June 20, 1961.
14. Interview with Hastings Keith, November 3, 1975.
15. Interview with Conrad Wirth, November 4, 1975.
16. Memorandum from Acting Assistant Director Robert W. Ludden to Director Conrad Wirth, undated.
17. Interview with Leo E. Diehl, November 3, 1975.
18. Letters requesting nominations sent by Secretary of the Interior Stewart Udall to boards of selectmen, September 29, 1961.
19. Press release, Department of the Interior, January 9, 1962.
20. Interview with Nathan Malchman, October 9, 1975.
21. Personal notes and/or recollections of the author.
22. Interview with George Palmer, October 10, 1975.
23. Charles H. W. Foster to Secretary of the Interior Stewart Udall, January 17, 1962.
24. *Boston Globe*, January 9, 1962.
25. Interview with Josiah H. Child, October 6, 1975.
26. Memorandum from Henry G. McCarthy to Charles H. W. Foster, undated.
27. *The Cape Codder*, January 4, 1962. Joshua A. Nickerson to Charles H. W. Foster, February 9, 1962.
28. Personal notes and/or recollections of the author.
29. Advisory Commission minutes, February 16, 1962.

30. Ibid.
31. Interview with Robert F. Gibbs, September 21, 1975.
32. *Cape Cod Standard-Times,* February 17, 1962.
33. *Boston Herald,* February 17, 1962.
34. Charles H. W. Foster to Regional Director Ronald F. Lee, February 19, 1962.
35. Interview with George H. Thompson, October 8, 1975.
36. Personal notes and/or recollections of the author.
37. Interview with Robert F. Gibbs, September 21, 1975.
38. Progress report from George H. Thompson to Director Conrad Wirth, January 22, 1962.
39. Interior field representative Mark Abelson to Interior resource policy staff director Charles H. Stoddard, March 30, 1962. Director Conrad Wirth to the Interior assistant secretary for fish, wildlife, and parks, August 7, 1962. Regional Director Ronald F. Lee to Superintendent Robert F. Gibbs, October 12, 1962.
40. Advisory Commission minutes, April 13, 1962.
41. Personal notes and/or recollections of the author.
42. Interview with Joshua A. Nickerson, October 8, 1975.
43. Personal notes and/or recollections of the author.
44. Executive Order 11007, President John F. Kennedy, February 9, 1962. Interior assistant secretary D. Otis Beasley to Director Conrad Wirth, June 7, 1962, confirming applicability to Advisory Commission.
45. Press release, Department of the Interior, April 16, 1962.
46. Advisory Commission minutes, March 9, 1962.
47. *The Cape Codder,* May 3, 1962. Interview with Conrad Wirth, November 4, 1975.
48. Interior under secretary John Carver to Secretary of the Army Elvis Stahr, March 16, 1962. Superintendent Robert F. Gibbs to Regional Director Ronald F. Lee, April 23, 1962. Commanding general, Fort Devens (Mass.), authorization for neutralization, May 23, 1962.
49. Advisory Commission minutes, March 9, 1962.
50. Advisory Commission minutes, January 6, 1967.
51. Advisory Commission minutes, April 13, 1962.
52. Regional Director Ronald F. Lee to Director Conrad Wirth, March 23, 1962.
53. Advisory Commission minutes, April 13, 1962.
54. Interview with George H. Thompson, October 8, 1975.
55. Interview with Robert A. McNeece, October 7, 1975; the "Great White Father" —a favorite phrase of Joshua A. Nickerson.
56. Regional Director Ronald F. Lee to Superintendent Robert F. Gibbs, June 5, 1962.
57. Interior secretary Stewart Udall to Charles H. W. Foster, November 15, 1962.
58. Director Conrad Wirth to Superintendent Robert F. Gibbs, July 23, 1962.
59. *Boston Herald,* October 14, 1962.
60. George H. Thompson to Charles H. W. Foster, March 15, 1962.
61. *New Beacon* (Provincetown, Mass.), January 10, 1962.
62. *The Cape Codder,* January 4, 1962. Joshua A. Nickerson to Governor John A. Volpe, January 5, 1962.
63. Interview with Josiah H. Child, October 6, 1975. *New Beacon* (Provincetown, Mass.), July 5, 1962.
64. Superintendent Robert F. Gibbs to Charles H. W. Foster, August 9, 1962. Director Conrad Wirth to Charles H. W. Foster, August 22, 1962.

65. Regional Director Ronald F. Lee to Director Conrad Wirth, March 23, 1962.
66. Charles H. W. Foster to Director of Planning Normand O. Pothier, Massachusetts Department of Commerce, January 22, 1962.
67. Interview with Elmer V. Buschman, November 5, 1975.
68. Superintendent Robert F. Gibbs to Regional Director Ronald F. Lee, November 2, 1962.
69. Except for Commission member Esther Wiles, see 1962 Wellfleet Town Report.
70. Advisory Commission minutes, January 11, 1963.
71. Advisory Commission minutes, September 27, 1963. Katharine Stone White, Minuteman National Historical Park Advisory Commission, to Charles H. W. Foster, September 30, 1963.
72. Superintendent Robert F. Gibbs to Charles H. W. Foster, September 30, 1963.
73. Advisory Commission minutes, November 1, 1963.
74. Advisory Commission minutes, April 26, 1963, and May 24, 1963.
75. Press release, Department of the Interior, June 10, 1963.
76. Massachusetts recreation planner Lewis A. Carter to Charles H. W. Foster, May 1, 1963.
77. Interview with Josiah H. Child, October 6, 1975. Josiah H. Child to Charles H. W. Foster: January 14, 1963; January 26, 1963; February 14, 1963; March 25, 1963; and August 21, 1963. Architects' responses: January 21, 1963 (Gropius); January 30, 1963 (Lawrence); February 5, 1963 (Gallagher); February 6, 1963 (Chermayeff); and February 8, 1963 (Belluschi).
78. Advisory Commission minutes, August 23, 1963.
79. *Cape Cod Standard-Times,* May 21, 1963. *Berkshire Eagle* (Pittsfield, Mass.), June 14, 1963. Interview with Vernon Gilbert, November 5, 1975.
80. Massachusetts director of forest and parks Raymond J. Kenney to Charles H. W. Foster, February 4, 1963. Superintendent Robert F. Gibbs to Charles H. W. Foster, March 5, 1963. Chief Ranger Von der Lippe to Charles H. W. Foster, October 7, 1963.
81. Advisory Commission minutes, January 11, 1963.
82. Ibid.
83. Ibid. Interview with Robert F. Gibbs, September 21, 1975.
84. George H. Thompson to Charles H. W. Foster, January 7, 1963. Charles H. W. Foster to George H. Thompson, March 15, 1963. Mrs. R. W. Werner, Great Beach Cottage Owners' Association, to Charles H. W. Foster, March 27, 1963. Superintendent Robert F. Gibbs to Regional Director Ronald F. Lee, June 27, 1963. Joshua A. Nickerson to Interior secretary Stewart Udall, July 11, 1963.
85. Regional Director Ronald F. Lee to Director Conrad Wirth, January 25, 1963. Telephone conversation, Regional Director Ronald F. Lee and Charles H. W. Foster, February 14, 1963. Superintendent Robert F. Gibbs to Charles H. W. Foster, May 14, 1963. Advisory Commission minutes, May 24, 1963.
86. Advisory Commission minutes, June 21, 1963.
87. Advisory Commission minutes, February 15, 1963, and September 27, 1963. Massachusetts commissioner of public works Ricciardi to Massachusetts state representative Reed, March 7, 1963. Charles H. W. Foster to Regional Director Ronald F. Lee, March 15, 1963. Regional Director Ronald F. Lee to Charles H. W. Foster, March 21, 1963.
88. Advisory Commission minutes, September 27, 1963.
89. Priscilla Redfield Roe, Fire Island National Seashore Advisory Commission, to Charles H. W. Foster, December 1963.

90. *State Journal* (Lansing, Mich.), December 15, 1963.
91. Joshua A. Nickerson, memorandum to Advisory Commission, April 16, 1964.
92. Advisory Commission minutes, June 25, 1965. Mrs. C. W. Post, petition against renaming the Seashore, December 2, 1963. Charles H. W. Foster to Interior secretary Stewart Udall, January 15, 1964. Museum archives correspondence (Eva Maria Dane): January 13, 1964; March 20, 1964; May 22, 1964; June 6, 1964; July 7, 1964; and July 17, 1964.
93. Federal Register, November 20, 1963. Regional Director Ronald F. Lee to Superintendent Robert F. Gibbs, April 23, 1964. Superintendent Robert F. Gibbs to U.S. representative Hastings Keith, July 9, 1964.
94. *Boston Herald,* May 3, 1964. Representative Hastings Keith to Charles H. W. Foster, October 28, 1964.
95. Superintendent Robert F. Gibbs, memorandum to Advisory Commission, August 14, 1963. Association for the Improvement of Medicine (AIM) correspondence (Adrian Murphy): January 2, 1964; February 12, 1964; February 28, 1964; March 7, 1964; and April 8, 1964.
96. Advisory Commission minutes: July 10, 1964; August 14, 1964; October 30, 1964; December 4, 1964; January 8, 1965; and February 19, 1965.
97. *Boston Globe,* October 18, 1964. Esther Wiles to Director Conrad Wirth, December 29, 1962.
98. Advisory Commission minutes, March 6, 1964.
99. Advisory Commission minutes, September 18, 1964.
100. Advisory Commission minutes, November 19, 1965. Judge M. Joseph Blumenfield, October 5, 1966. Interior assistant secretary Stanley A. Cain to Wilfred Svenson, Massachusetts Council of Sportsmens Clubs: Great Island to remain a natural area; undated. Esther Wiles, 1966 Wellfleet Town Report.
101. Advisory Commission minutes, November 19, 1965. Massachusetts congressional delegation, joint letter to National Park Service director George F. Hartzog, Jr., May 7, 1965; his response, June 24, 1965. Joshua A. Nickerson correspondence: June 7, 1965 (Keith); July 4, 1965 (Keith); July 6, 1965 (Keith); and July 8, 1965 (Saltonstall). Director George F. Hartzog, Jr., to Charles H. W. Foster, October 21, 1965.
102. Advisory Commission minutes, May 20, 1966.
103. Regional Director Ronald F. Lee to Charles H. W. Foster, January 11, 1966. *Boston Globe,* August 6, 1966. Charles H. W. Foster to Interior secretary Stewart Udall, November 21, 1966.
104. Director George F. Hartzog, Jr., to Charles H. W. Foster, January 4, 1966.
105. Advisory Commission minutes, February 25, 1966.
106. Regional Director Ronald F. Lee to Charles H. W. Foster, January 11, 1966.
107. Interview with Esther Wiles, October 21, 1975.
108. Superintendent Stanley C. Joseph to Regional Director Lemuel A. Garrison, August 9, 1966. Telephone discussions, Superintendent Stanley C. Joseph, Regional Director Lemuel A. Garrison, and Charles H. W. Foster, July 22, 1966, and August 26, 1966.
109. Advisory Commission minutes, September 23, 1966.
110. *Boston Globe* and *Boston Herald,* August 6, 1966.
111. Advisory Commission minutes, January 14, 1966.
112. Advisory Commission minutes, August 26, 1966.
113. Truro Golf Committee to Superintendent Stanley C. Joseph, February 7, 1967.
114. Advisory Commission minutes, May 20, 1966.

115. Josiah H. Child to Charles H. W. Foster, October 30, 1966.
116. Advisory Commission minutes, April 15, 1966.
117. Esther Wiles, 1966 Wellfleet Town Report.
118. Advisory Commission minutes, December 2, 1966.
119. New Bedford *Standard Times*, December 17, 1969.
120. Cape Cod Chamber of Commerce (Norman Cook) to Joshua A. Nickerson, October 25, 1967.
121. Advisory Commission minutes, November 17, 1967. Cape Cod *Standard-Times*, July 25, 1974.
122. Esther Wiles, 1967 Wellfleet Town Report. Lands Exchange Committee (Wellfleet) to Superintendent Stanley C. Joseph, October 5, 1967. Superintendent Stanley C. Joseph to Regional Director Lemuel A. Garrison, October 6, 1967. Regional Director Lemuel A. Garrison to Superintendent Stanley C. Joseph, November 9, 1967.
123. Advisory Commission minutes, August 2, 1968.
124. Charles H. W. Foster to Josiah H. Child, October 24, 1966.
125. Josiah H. Child to Norton H. Nickerson, August 9, 1968. Governor Francis W. Sargent to Interior secretary Walter Hickel, November 17, 1969. Interview with Norton H. Nickerson, May 24, 1984.
126. Regional Director Lemuel A. Garrison to Director George F. Hartzog, Jr., November 25, 1968.
127. Superintendent Stanley C. Joseph, memorandum to Advisory Commission, February 21, 1968.
128. Advisory Commission minutes, September 27, 1968.
129. Joshua A. Nickerson's term; Advisory Commission minutes, February 28, 1969.
130. Except for Esther Wiles, 1968 Wellfleet Town Report: "gets children out of the home and weakens the control of the parents over them"; leads to "enslavement . . . destruction of morality."
131. Nauset Regional School District to Superintendent Leslie P. Arnberger, November 13, 1969.
132. Advisory Commission minutes, April 17, 1970.
133. Superintendent Leslie P. Arnberger to regional director, November 25, 1970.
134. Press release, Department of the Interior, May 31, 1970.
135. Leo E. Diehl to Superintendent Leslie P. Arnberger (handwritten note), July 29, 1970. Interview with Leo E. Diehl, November 3, 1975.
136. Advisory Commission minutes, September 11, 1970.
137. Interior assistant solicitor David A. Watts to director, undated.
138. Superintendent Leslie P. Arnberger, memorandum to Advisory Commission, November 20, 1970.
139. Superintendent Leslie P. Arnberger to files, report of meeting with state and town officials, March 12, 1971.
140. Joshua A. Nickerson to U.S. senator Henry Jackson, November 24, 1971. Joshua A. Nickerson, letter to the editor, *Vineyard Gazette*, April 25, 1972. Joshua A. Nickerson to Superintendent Leslie P. Arnberger (telephone), January 20, 1972.
141. Public Law 92-463, October 6, 1972.
142. Federal Register, June 7, 1972.
143. Regional director to all superintendents, November 24, 1972.
144. Regional director to all superintendents, call for justification statements for all

existing advisory commissions, October 11, 1972. Response from Superintendent Leslie P. Arnberger, October 27, 1972: "advice . . . most helpful"; "continuation . . . would be desirable." Superintendent Leslie P. Arnberger to Regional Director Chester Brooks, November 1, 1973. Deputy Regional Director David A. Richie to Regional Director Jerry Wagers, July 11, 1974.

145. An appointment urged by Robert A. McNeece (interview with David Ryder, May 14, 1984).
146. Advisory Commission minutes, February 23, 1973.
147. Advisory Commission minutes, September 21, 1973.
148. Advisory Commission resolution, February 22, 1974.
149. Interview with Lawrence C. Hadley, October 24, 1975. Press release, Department of the Interior, December 23, 1973.
150. Advisory Commission minutes, February 23, 1974.
151. Superintendent Lawrence C. Hadley to Yosemite National Park superintendent Leslie P. Arnberger, March 6, 1974.
152. Superintendent Lawrence C. Hadley to Joshua A. Nickerson, May 23, 1974.
153. Press release, Association for the Preservation of Cape Cod, March 8, 1974. Superintendent Lawrence C. Hadley to Massachusetts commissioner of natural resources Arthur W. Brownell, April 16, 1974.
154. Advisory Commission minutes, September 20, 1974.
155. Advisory Commission minutes, November 22, 1974.
156. Statement of Advisory Commission, November 22, 1974.
157. Superintendent Lawrence C. Hadley to regional director, December 31, 1974.
158. Interview with Esther Wiles, October 21, 1975.
159. Wellfleet selectman Plante to secretary of the interior, August 27, 1974.
160. Superintendent Lawrence C. Hadley to Esther Wiles, September 30, 1974.
161. *Provincetown Advocate,* September 12, 1974.
162. Response from Eastham (March 7, 1975); Orleans (March 7, 1975); Truro (March 14, 1975); Wellfleet (April 15, 1975); Provincetown (April 16, 1975); Chatham (March 23, 1977); Barnstable County (April 23, 1975); Commonwealth of Massachusetts (August 12, 1975).
163. Advisory Commission minutes, May 9, 1975.
164. Robert Landau to Superintendent Lawrence C. Hadley, May 13, 1975. Interview with Robert Landau, October 17, 1975. Interview with Shirley Luikens, March 23, 1984.
165. *The Cape Codder,* July 31, 1975. Advisory Commission minutes, September 5, 1975.
166. Advisory Commission minutes, July 11, 1975.
167. Advisory Commission minutes, September 5, 1975.
168. Advisory Commission minutes, January 23, 1976.
169. Advisory Commission minutes, July 11, 1975.
170. *The Cape Codder,* July 17, 1975.
171. Advisory Commission minutes, April 16, 1976.
172. Advisory Commission minutes, January 23, 1976. *Provincetown Advocate,* February 5, 1976: "When is a house not a house but a different house?"
173. Advisory Commission minutes, April 16, 1976. *The Cape Codder,* April 22, 1976: "Face it, we need to find oil."
174. Advisory Commission minutes, October 8, 1976.
175. Martin W. Peters to Joshua A. Nickerson, March 16, 1976.
176. *The Cape Codder,* January 27, 1977.

177. Dexter M. Keezer to Senator Edward Brooke, April 18, 1977.
178. Advisory Commission minutes, April 1, 1977.
179. Advisory Commission minutes, June 17, 1977.
180. *Cape Cod Times,* January 26, 1976.
181. Advisory Commission minutes, June 17, 1977.
182. Advisory Commission minutes, August 19, 1977.
183. Advisory Commission minutes, October 14, 1977. For public commentary on this issue, see series of articles in *The Cape Codder* (October 14, 1976 and January 6, 1977), reprinted in the Congressional Record, January 25, 1977. *Provincetown Advocate,* November 18, 1976; "The Jane, rebuilt piece by piece, is still the Jane."
184. Clair L. Baisly, Chatham Historical Commission, to Superintendent Lawrence C. Hadley, December 16, 1977.
185. Advisory Commission minutes, October 14, 1977.
186. Advisory Commission minutes, December 9, 1977.
187. Ibid.
188. Ibid.
189. *The Cape Codder,* January 10, 1978. *Cape Cod Times,* January 10, 1978, January 19, 1978. *The Cape Codder,* editorial, March 10, 1978. Joshua A. Nickerson to Superintendent Lawrence C. Hadley, January 12, 1978, and July 18, 1978. Orleans selectman Gaston Norgeot to Speaker Thomas P. O'Neill, undated.
190. Advisory Commission minutes, October 14, 1977.
191. *The Cape Codder,* April 7, 1978.
192. Records of the Office of Advisory Boards and Commissions, National Park Service, Washington, D.C. Interview with Shirley Luikens, March 23, 1984.
193. The term used by Robert Landau in correspondence with Superintendent Lawrence C. Hadley, May 13, 1975.
194. See Appendix F for current charter containing such language.
195. Advisory Commission minutes, August 25, 1978.
196. Joshua A. Nickerson to Superintendent Lawrence C. Hadley, July 28, 1978: "not a candidate for chairman." Interview with David F. Ryder, May 14, 1984. Dexter M. Keezer to Joshua A. Nickerson, January 16, 1978.
197. Advisory Commission minutes, November 3, 1978.
198. Ibid.
199. Interview with Superintendent Herbert Olsen, May 14, 1984.
200. Advisory Commission minutes, November 3, 1978.
201. Interview with Sally H. Lunt, May 17, 1984.
202. Interview with Clifford H. White, May 16, 1984.
203. Advisory Commission minutes, March 23, 1979.
204. Ibid.
205. Ibid.
206. Advisory Commission minutes, June 1, 1979.
207. Advisory Commission minutes, October 19, 1979.
208. Advisory Commission minutes, December 14, 1979. Interview with Barbara S. Mayo, May 15, 1984.
209. Advisory Commission minutes, December 14, 1979.
210. Ibid.
211. Advisory Commission minutes, March 7, 1980.
212. *The Cape Codder,* letter to the editor, Elizabeth Foss-Mayo (Tranquility Lobby), May 16, 1980.

213. Advisory Commission minutes, March 7, 1980.
214. Land Acquisition Plan, Cape Cod National Seashore, approved March 27, 1980. Interview with James Killian, May 14, 1984.
215. Advisory Commission minutes, March 7, 1980.
216. Advisory Commission minutes, May 2, 1980.
217. Advisory Commission minutes, July 11, 1980.
218. Advisory Commission minutes, September 12, 1980.
219. Ibid.
220. Advisory Commission minutes, November 14, 1980.
221. Advisory Commission minutes, January 16, 1981.
222. Ibid.
223. Advisory Commission minutes, April 24, 1981.
224. Advisory Commission minutes, June 12, 1981.
225. Advisory Commission minutes, August 7, 1981. Interview with Superintendent Herbert Olsen, May 14, 1984.
226. Thoreau, *Cape Cod*, [1864] 1961.
227. Interview with Jonathan Moore, October 29, 1975.
228. Interview with Leo E. Diehl, November 3, 1975.
229. Interview with Frederick Holborn, November 4, 1975.
230. Interview with Conrad Wirth, November 4, 1975. Interview with Ben H. Thompson, November 4, 1975.
231. Interview with George Palmer, October 10, 1975.
232. Robert F. Gibbs, "Cape Cod National Seashore Case Study," November 1975 (unpublished).
233. Interview with George H. Thompson, October 8, 1975.
234. Winford L. Schofield, "National Seashore Survey Formula and Idle Prattle," Park Service files, memo, September 4, 1962.
235. Comment made by Ralph Chase (Eastham), Advisory Commission minutes, February 9, 1968.
236. Interview with David B. H. Martin, November 3, 1975.
237. Report of Subcommittee on Land Acquisition, Advisory Commission minutes, March 7, 1980.
238. Interview with Elmer V. Buschman, November 5, 1975.
239. Report of Subcommittee on Guidelines/Criteria for Private Property Development, Advisory Commission minutes, October 14, 1977.
240. Interview with Elmer V. Buschman, November 5, 1975.
241. Public Law 94-565 (1976) as described in Advisory Commission minutes, October 8, 1982.
242. See Appendix E.
243. Senate Report No. 428, June 20, 1961.
244. Letter from Eastham Board of Selectmen, March 7, 1975.
245. Letter from Wellfleet Board of Selectmen, April 15, 1975.
246. Ibid.
247. Ibid.
248. Letter from Truro Board of Selectmen, March 14, 1975.
249. Robert Landau to Superintendent Lawrence C. Hadley, May 13, 1975.
250. Malcolm Dickinson (Orleans Conservation Commission) and Warrenton Williams (Cape Cod Planning and Economic Development Commission), Advisory Commission minutes, April 17, 1970.
251. Eastham hearing, Senate Subcommittee on Public Lands, December 9, 1959.

252. Visit to Cape Cod on October 3–4, 1968.
253. Visit to Concord on September 27, 1963.
254. Interview with Robert F. Gibbs, September 21, 1975.
255. Letter from Stanley C. Joseph, October 28, 1975.
256. Letter from Leslie P. Arnberger, December 2, 1975.
257. Interview with Lawrence C. Hadley, October 24, 1975.
258. Interview with Superintendent Herbert Olsen, May 14, 1984.
259. Ronald F. Lee to Charles H. W. Foster, January 11, 1966.
260. Interview with Dexter M. Keezer, May 23, 1984.
261. Personal notes and/or recollections of the author.
262. Interview with David F. Ryder, May 14, 1984.
263. Interview with Clifford H. White, May 16, 1984.
264. Composite of interviews with sixteen past and present Advisory Commission members.
265. Superintendent Lawrence C. Hadley to Esther Wiles, September 30, 1974.
266. Interview with Esther Wiles, October 21, 1975.
267. Ibid.
268. Advisory Commission minutes, September 20, 1974.
269. Esther Wiles to Secretary of the Interior Rogers Morton, January 9, 1975.
270. Interview with Sally H. Lunt, May 17, 1984.
271. Joshua A. Nickerson, correspondence with Chairman Carl T. Johnson, Sleeping Bear Dunes National Lakeshore, 1971.
272. Eastham hearing, Senate Subcommittee on Public Lands, December 9, 1959.
273. Ibid.
274. Interview with Jonathan Moore, October 29, 1975.
275. Interview with Conrad Wirth, November 4, 1975.
276. Senate Report No. 402, June 20, 1961.
277. Interview with Norton H. Nickerson, May 24, 1984.
278. National Park Service, *State of the Parks 1980* (Report to Congress), May 1980.
279. Interview with Dexter M. Keezer, May 23, 1984.

Selected References

REPORTS AND PUBLISHED WORKS

Advisory Commission on Intergovernmental Relations. 1972. *Multistate regionalism.* Washington, D.C.: Government Printing Office.

Beston, Henry. [1928] 1983. *The outermost house.* Reprint. New York: Penguin Books.

Beyle, Thad Lewis. 1963. The Cape Cod National Seashore: A study in conflict. Ph.D. diss., University of Illinois, Urbana-Champaign.

Burling, Francis P. 1978. *The birth of the Cape Cod National Seashore.* Plymouth, Mass.: Leyden Press.

Damore, Leo. 1967. *The Cape Cod years of John Fitzgerald Kennedy.* Englewood Cliffs, N.J.: Prentice Hall.

Fearson, Robert H. 1977. *The Cape Cod Canal.* Middletown, Conn.: Wesleyan University Press.

Freeman, Frederick. 1862. *The history of Cape Cod: The annals of the thirteen towns of Barnstable County, including the district of Mashpee.* Boston: Geo. C. Rand & Avery.

Giambarba, Paul. 1968. *Cape Cod and Cape Cod National Seashore.* Centerville, Mass.: Scrimshaw Press.

Hay, John. 1972. *The great beach.* New York: Ballantine Books.

Kittredge, Henry C. [1930] 1968. *Cape Cod: Its people and their history.* Reprint. Boston and New York: Houghton Mifflin Co.

Massachusetts Department of Labor and Industries. 1922. *Population and resources of Cape Cod: A special report in recognition of the three hundredth anniversary of the settlement of New England.* Boston: Wright & Potter Printing Co.

Smith, Kenneth A. 1976. Citizen advisors and national parks. Report on a study for the National Park Service (Midwest Region). Omaha, Neb.

Strahler, Arthur N. 1966. *A geologist's view of Cape Cod.* Garden City, N.Y.: Natural History Press.

Thoreau, Henry David. [1864] 1961. *Cape Cod.* Reprint. New York: Thomas Y. Crowell Co.

U.S. Department of the Interior. National Park Service. 1956. *Our vanishing shoreline.* Washington, D.C.: Government Printing Office.

————. 1959. *Cape Cod National Seashore: A proposal.* Washington, D.C.: Government Printing Office.

CONGRESSIONAL DOCUMENTS

Hearings on S. 2636 (a bill to provide for the establishment of Cape Cod National Seashore Park), December 9 and 10, 1959. Subcommittee on Public Lands, Senate Committee on Interior and Insular Affairs. Eastham, Mass.

Hearings on S. 2636, June 21, 1960. Subcommittee on Public Lands, Senate Committee on Interior and Insular Affairs. Washington, D.C.

Hearings on various proposals for the establishment of the Cape Cod National Seashore Park in the Commonwealth of Massachusetts, December 16 and 17, 1960.

Subcommittee on Public Lands, House Committee on Interior and Insular Affairs. Eastham, Mass.

Hearings on H.R. 989 and related bills to provide for establishment of Cape Cod National Seashore Park, March 6, 7, and 8, 1961. House Committee on Interior and Insular Affairs. Washington, D.C.

Hearings on S. 857 to provide for the establishment of Cape Cod National Seashore Park, March 9, 1961. Subcommittee on Public Lands, Senate Committee on Interior and Insular Affairs. Washington, D.C.

Senate Report 428, Senate Committee on Interior and Insular Affairs, June 20, 1961. Providing for establishment of Cape Cod National Seashore Park.

House Report 673, House Committee on Interior and Insular Affairs, July 3, 1961. Establishing Cape Cod National Seashore, Mass.

House Report 831, Conference Committee Report to accompany S. 857, August 1, 1961. Establishment of Cape Cod National Seashore.

Public Law 87-126, August 7, 1961. Establishment of Cape Cod National Seashore.

Public Law 91-252, May 14, 1970. Increase to $33,500,000, authorization for purchase of land within Cape Cod National Seashore.

Public Law 92-463, October 6, 1972. Federal Advisory Committee Act.

Public Law 98-141, October 31, 1983. Increase to $42,917,575, authorization for purchase of land within Cape Cod National Seashore.

Index